"十二五"国家重点出版物出版规划项目

本书得到下列项目资助：
国家自然科学基金项目(41171053)
冰冻圈科学国家重点实验室自主课题

英汉冰冻圈科学词汇

English-Chinese Dictionary of Cryospheric Sciences

主　编：秦大河
副主编：姚檀栋　丁永建　任贾文

China Meteorological Press

内容简介

本书是第一版冰冻圈科学词汇，共收录 8448 个词条，涉及冰冻圈科学领域常见的英文词汇及其对应的中文，可供冰冻圈科学及地理、水文、地质地貌、大气、生态、环境、海洋等方面的科研和技术人员，以及大专院校相关专业的师生使用和参考。

图书在版编目(CIP)数据

英汉冰冻圈科学词汇／秦大河等主编．—北京：气象出版社，2012.8
ISBN 978-7-5029-5549-6

Ⅰ.①英… Ⅱ.①秦… Ⅲ.①冰川学－词汇－英、汉
Ⅳ.①P343.6-61

中国版本图书馆 CIP 数据核字(2012)第 194946 号

英汉冰冻圈科学词汇

出版发行：	气象出版社			
地　　址：	北京市海淀区中关村南大街 46 号	邮政编码：	100081
网　　址：	http://www.cmp.cma.gov.cn	E-mail：	qxcbs@cma.gov.cn
总 编 室：	010-68407112	发 行 部：	010-68409198
责任编辑：	蔺学东	终　　审：	周诗健
封面设计：	博雅思企划	责任技编：	吴庭芳
印　　刷：	北京中新伟业印刷有限公司		
开　　本：	889 mm×1194 mm　1/32	印　　张：	8.5
字　　数：	330 千字		
版　　次：	2012 年 9 月第 1 版	印　　次：	2012 年 9 月第 1 次印刷
定　　价：	45.00 元		

本书如存在文字不清、漏印以及缺页、倒页、脱页者，请与本社发行部联系调换。

序　　言

　　过去数十年来,全球气候经历了以变暖为主要特征的显著变化。在地球表层系统中,冰冻圈"感知"这一变化最为快速,并对全球和区域气候系统产生了强烈反馈作用。这也催生了近年来冰冻圈研究逐渐以独立学科出现,改变了过去将冰冻圈简单地作为水圈一部分的传统认识。这一转变首先是20世纪80年代气候系统科学概念提出后,将冰冻圈作为独立圈层,标志性事件则是2007年新成立的国际冰冻圈科学协会(IACS),国内则以2007年组建冰冻圈科学国家重点实验室为标志。

　　然而,由于冰冻圈科学与其他学科的高度交叉性,将冰冻圈视为整体性学科体系后对其内涵、外延的界定需要从理论体系和应用领域方面重新梳理,国际上尚无现成的线索和模版可供参考。好在,中国几代科学家经历了半个多世纪的研究积累,对冰冻圈各要素的研究较为全面,具备探索并率先形成冰冻圈科学框架的优势条件。要构建这样一座新的学科大厦绝非易事,首先要从一砖一瓦做起,而最基本的"砖瓦"之一就是统一冰冻圈科学的学术术语。以秦大河院士为首的我国冰冻圈科学群体,邀请大气圈、生物圈、水圈、岩石圈表面和社会人文科学相关领域的科学家,集思广益,编纂了首部《英汉冰冻圈科学词汇》,这是我国冰冻圈科学研究历程中的一件大事,也将为国际科学界提供重要参考,具有深远意义。

为此,我祝贺《英汉冰冻圈科学词汇》的出版,也希望编写组再接再厉,不断推出本领域辞典、教科书,乃至百科全书的学科单本,为指导学科发展,并为推动与冰冻圈科学相关的经济社会发展作出更大贡献。

2012 年 2 月 29 日

前　言

　　冰冻圈(cryosphere)系指地球表面(包括陆地和海洋表面)之上和之下一定范围内以冻结状态存在的水体及其混合物，包括北冰洋底下伏的多年冻土，也包括大气圈内冻结状态的水体。冰冻圈的主要组成部分为冰川、冰盖(格陵兰冰盖和南极冰盖)、冻土(季节冻土和多年冻土)、积雪、江、河、湖、海冰和冰架。目前，全球地表约75%的淡水资源储存在冰冻圈中，陆地表面约10%被现代冰川覆盖，海冰面积约占海洋面积的7%，北半球冬季积雪面积为陆地面积的49%，冻土面积约占陆地面积的四分之三，而多年冻土覆盖了北半球陆地面积的24%(IPCC WG1，2007)。

　　冰冻圈对气候高度敏感，并具有反馈作用，是气候系统五大圈层之一。冰冻圈也对自然环境和人类社会产生重要影响，与可持续发展息息相关，在全球变暖的今天其重要性日趋显著。在全球气候变化研究中，冰冻圈以其巨大体量所具有的冷储、相变潜热，广袤面积所具备的反照率，以及温室气体，特别是碳的源、汇及其转换等，使其重要性在气候系统五大圈层中仅次于海洋(水圈)。此外，发掘冰冻圈内的地球环境气候演变记录，可以重建约百万年以来地球气候环境演化旋回和变化细节，解疑全球变化研究中的难点和疑点。在世界许多地区，冰冻圈是社会经济发展所需的水源地,也是干旱区生态系统得以发育和保护的屏障。例如,发源于亚洲内陆腹地和青藏高原冰冻圈的众多河流里,印度河、恒河和雅鲁藏布江最终注入印度洋,黄河和长江向东流入太平洋,而向北有鄂毕河进入北冰洋,沿途滋养哺育着近30亿人口,占世界总人口的40%以上。

与冰冻圈学科相关的科学和生产实践在国内外已有很长的历史。随着社会发展和科学进步,特别是20世纪末到本世纪初,全球气候变化问题的突显,与冰冻圈相关的种种问题受到学术界和社会各界前所未有的关注,使冰冻圈科学成为地球科学中发展最迅速、最活跃的领域之一。由于冰冻圈科学是一门交叉学科,与大气、地理、地质、水文、生态、海洋、物理、化学等许多学科密切关联,同时还需要运用测量、遥感、数值模型、地理信息系统及实验分析等技术手段,而且国际性又很强,相关的科学术语及其中英文对照等存在较大的问题,妨碍了学科之间交流和国际交流,也容易引起误解。同一个词往往被用来表述不同的含义,而同一个被描述的对象却可能有不同的名称,尤其是专业英语词汇的中文翻译,存在较大的混乱。例如,英文中的 snow cover,本意是指沉降积存在地面上的雪,其存留时间不超过12个月,实际是指"季节性积雪",简称"积雪"。但有些读者望文生义,将其译成"雪盖",完全抛弃了英文原意中的时间概念。还有的仍在使用20世纪之前的古老译法,不符合现代科学语言,如将冰架(ice shelf)译为陆缘冰,将粒雪(firn)译为万年雪或永久雪;或移植日语的译法,将冰川(glacier)译为冰河;还有将多年冻土(permafrost)译为永久冻土,殊不知在全球变暖的今天,多年冻土不可能一成不变地、"永久"地冻结。还有将冰盖(ice sheet)译成冰层、冰原,等等,造成混乱和概念混淆,对科学本身和科学普及都带来损失。

影响扩大,学科交叉,社会关注,许多学科和社会各界关注并使用冰冻圈科学的研究成果,所以统一冰冻圈科学中的有关术语和词汇的中英文对照,既很必要,也很迫切。2007年,冰冻圈科学国家重点实验室的建立标志着我国冰冻圈科学体系的形成,尽早编纂出版有关冰冻圈科学词汇和辞典也逐渐达成共识。自2010年8月,《英汉冰冻圈科学词汇》的编纂工作历时一年半,期间召开了八次编辑会议,至2012年2月最终定稿。最初提供备选词条供挑选的人员有60多人,其中30多人参加了编辑讨论和评审。由于是首次编辑《英汉冰冻圈科学词汇》,经验不足,时间紧迫,加上涉及的学科面广,某些词汇肯定有遗漏,有的中英文对应不准确。希望有关专家和使用者能够批评指正,以便再版时补充或修改。

编　者

2012年6月

顾问委员会

(按姓氏拼音排序)

安芷生	巢纪平	陈宜瑜	程国栋	丁一汇	丁仲礼
傅伯杰	郭华东	胡敦欣	李吉均	李家洋	李文华
李小文	刘昌明	穆　穆	石广玉	施雅风	苏纪兰
孙鸿烈	陶　澍	吴国雄	徐冠华	袁道先	张建云
张新时	郑　度	周卫健	周秀骥		

编委会名单

主　编：秦大河
副主编：姚檀栋　丁永建　任贾文

主要编纂和评审人员（按姓氏拼音排序）：
车　涛　陈肖柏　丁永建　董治宝　方一平　何元庆
侯书贵　金会军　康世昌　赖远明　李培基　李　新
李忠勤　刘时银　罗　勇　潘保田　钱正安　秦大河
任贾文　施建成　孙　波　孙俊英　王宁练　王秋凤
吴青柏　武炳义　效存德　杨大庆　杨建平　姚檀栋
叶柏生　于贵瑞　俞卫平　张人禾　张廷军　赵　林
赵井东　周尚哲

秘书组：
王文华　宜树华　马丽娟　谢爱红　王亚伟　明　镜
俞　杰

提供备选词条人员（按姓氏拼音排序）：
卞林根　曹建廷　车　涛　陈　拓　陈肖柏　丁永建
董文杰　董治宝　方一平　高学杰　侯书贵　金会军

康世昌　赖远明　李培基　李世杰　李　新　李志军
李忠勤　刘时银　罗　勇　马　巍　马耀明　潘保田
钱正安　秦大河　任贾文　施建成　孙　波　孙俊英
王根绪　王国庆　王宁练　王秋凤　魏文寿　吴青柏
武炳义　效存德　杨大庆　杨建平　杨兴国　姚檀栋
叶柏生　于贵瑞　俞卫平　张东启　张人禾　张廷军
张小曳　张耀南　赵　林　赵井东　赵　平　周尚哲
朱　诚

特别感谢孙鸿烈院士为本书作序。

使用说明

一、全部英文词汇按英文字母顺序排列。

二、英文词汇后圆括号中,如有等号,等号后面的词是同义词。

如:barchan (= barchane = barkhan) 新月形沙丘,新月形雪堆,即:barchans = barchane = barkhan。

三、英文词汇后圆括号中,如没有等号,则括号内的是缩略词。

如:snow water equivalent (SWE) 雪水当量。

四、英文词汇后圆括号中,如没有等号,并且是斜体字,则括号内的是拉丁语。

如:Tibetan antelope (*Pantholops hodgsoni*) 藏羚羊。

五、英文词汇后方括号的部分可以省略。

如:ice lead[er] 冰间水道,水沟。

六、中文解释中加圆括号进行修饰、提供地名或语言出处。

如:pluck (冰川)拔蚀、拔削;

Vostock ice core 东方站冰芯(南极);

Schneebrett 雪崩(德语)。

七、中文解释中加方括号,其中内容在应用时可以省略。

如:ablation 消融[作用]。

目 录

序言

前言

使用说明

英汉冰冻圈科学词汇 ·································· (1—165)

附录1 地质年表 ·································· (167)

附录2 国内外著名冰川 ·································· (169)

附录3 国内外主要期刊 ·································· (186)

　　　(1) 英文期刊 ·································· (186)

　　　(2) 中文期刊 ·································· (192)

附录4 国内外主要研究机构 ·································· (198)

　　　(1) 国际冰冻圈研究机构 ·································· (198)

　　　(2) 国内冰冻圈相关研究机构 ·································· (207)

附录5 冰冻圈区主要科学考察站 ·································· (210)

　　　(1) 南极地区的冰冻圈研究站 ·································· (210)

　　　(2) 北极地区的冰冻圈研究站 ·································· (217)

　　　(3) 其他地区的冰冻圈研究站 ·································· (221)

附录6 主要学术组织(冰冻圈相关) ·································· (226)

附录7 主要科学计划(冰冻圈相关) ·································· (228)

附录8 与冰冻圈相关的主要缩略词 ·································· (232)

英汉冰冻圈科学词汇

A

ablation 消融〔作用〕
ablation area (zone) 消融区
ablation cone 冰锥
ablation drift 消融冰碛
ablation intensity 消融强度
ablation model 消融模型
ablation moraine 消融冰碛垄
ablation period 消融期
ablation rate 消融速率
ablation till 消融冰碛
ablatograph 冰融仪
ablatography 消融过程线
above sea level (a.s.l.) 海拔
ab-polar current 离极气流
abrasion （冰川）磨蚀
abrupt climate change 气候突变
abrupt climate event 气候突发事件
absolute accuracy 绝对精度
absolute age 绝对年龄
absolute altimeter 绝对测高计
absolute altitude 绝对高度,绝对高程
absolute dating 绝对年代测定
absolute elevation 绝对高程
absolute flying height 绝对航高
absolute gravimeter 绝对重力仪
absolute gravity measurement 绝对重力测量
absolute growth rate 绝对生（增）长率
absolute stability 绝对稳定性
absolute temperature 绝对温度
absolute thermometer 绝对温度计
absolute water content 绝对含水量
absorbability 吸收性
absorbance 吸收度
absorbed dose 吸收剂量
absorbed layer 吸附层
absorbing medium 吸收介质
absorption 吸收作用
absorption coefficient 吸收系数
absorption constant 吸收常数
absorption cross section 吸收截面
absorption factor 吸收因子
absorption index 吸收指数
absorption spectrum 吸收光谱
absorptivity 吸收率
absorptivity-emissivity 吸收发射率
abundance 丰度
abundance zone 富集带
abyssal circulation 深海环流,深层环流
abyssal facies 深海相
abyssal sediment 深海沉积物
Acadian 阿卡德统（中寒武世）
accelerated erosion 加速侵蚀

accelerator mass spectrometry (AMS) 加速器质谱仪
acceptable erosion 容许侵蚀量
accessory species 次要种
accidental species 偶见种
acclimatization 环境适应,气候适应
accreted ice 增生冰
accretion 堆积,冲积,碰并增长(云物理)
accretion efficiency 碰并效率
accretion moraine 前进终碛垄
accumulated deformation 累积变形
accumulated displacement 累积位移
accumulated error 累积误差
accumulated plastic strain 累积塑性应变
accumulated snowdrift area 风吹雪堆积区
accumulating temperature 积温
accumulation 积累〔作用〕,堆积〔作用〕
accumulation area (=zone) 积累区
accumulation area ratio (AAR) 积累区面积比率
accumulation horizon 堆积层
accumulation landform 堆积地貌
accumulation period 积累期,沉积期
accumulation platform 堆积台地
accumulation rate 积累率
accumulation relief 堆积地形
accumulation terrace 堆积阶地
accumulative mass balance 累积物质平衡
accumulative precipitation gauge 总降水器(计)
accumulative raingauge 总雨量器(计)
acicular ice 屑冰,针状冰,纤维状冰
acicular-leaved tree 针叶树
aciculifruticeta 针叶灌木群落
aciculignosa 针叶木本群落
acid 酸
acid deposition 酸沉降
acid precipitation 酸性降水
acid rain 酸雨
acid-base titration 酸碱滴定法
acidification 酸化〔作用〕
acidify 酸化
acoustic depth telemeter 声学测深仪
acoustic holography system 水声全息系统
acoustic log 声波测井
acoustic ocean-current meter 声学海流计
acoustic positioning 水声定位,声波定位
acoustic water level 声学水位计
acrocyanosis 冻疮
activation energy 活化能
activation free energy 活化自由能
active aurora 活动极光
active fault 活断层

active ice wedge 活动冰楔
active layer 活动层
active layer detachment 活动层边坡坍塌
active layer failure 活动层破坏
active layer thickness 活动层厚度
active melt period 高速解冻期
active microwave instrument 主动微波传感器
active moraine 活动冰碛
active rock glacier 活动石冰川
active sensor 主动遥感器
active volcano 活火山
activity 活化度,活性
actual evapotranspiration 实际蒸散〔量〕
actual load 有效载荷
actual stress 实际应力
adaptability 适应性
adaptation 适应
adaptation policy 适应政策
adaptation process 适应过程
adaptation time 适应时间
adaptive algorithm 自适应算法
adaptive management 适应性管理
adaptive selection 适应性选择
adaptive strategy 适应策略
adfluxion 汇流
adfreeze 附加冻结作用
adfreezing force 冻结力
adfreezing ice force 附加冰冻结力
adfreezing strength 附加冻结强度
adiabatic 绝热,隔热
adiabatic ascending 绝热上升
adiabatic cooling 绝热冷却
adiabatic heating 绝热增温
adiabatic lapse rate 绝热直减率
adiabatic process 绝热过程
adiabatic sinking 绝热下沉
adret 阳坡
adsorption 吸附〔作用〕
Advanced Along-Track Scanning Radiometer（AATSR） 高级距离向扫描辐射计
Advanced Earth Observation Satellite（ADEOS） 先进对地观测卫星
Advanced Earth Resources Observation System（AEROS） 高级地球资源观测系统
Advanced Land Observation System（ALOS） 先进陆地观测系统
Advanced Microwave Scanning Radiometer（AMSR） 高级微波扫描辐射计
Advanced Microwave Scanning Radiometer for Earth Observing System（AMSR-E） 地球观测系统高级微波扫描辐射计
Advanced Microwave Sounding Unit（AMSU） 高级微波探测器
Advanced Spaceborne Thermal Emission and Reflection Radiometer（ASTER） 先进型星载热发射反射辐射计
Advanced Synthetic Aperture Radar

（ASAR） 先进合成孔径雷达
Advanced Topographic Laser Altimeter System（ATLAS） 高级地形激光测高系统
Advanced Very High Resolution Radiometer（AVHRR） 改进型超高分辨率辐射计
Advanced Visible and Near Infrared Radiometer type 2（AVNIR-2） 高级可见与近红外辐射计（二型）
advancing glacier 前进冰川
advection 平流
advection transfer 平流输送
advection-diffusion equation 平流－扩散方程
adventitious species 外来种，侵入种
adverse pressure gradient 逆压梯度
adversity selection 逆境选择
aeolian 风成的
aeolian accumulation 风成堆积〔作用〕
aeolian deposit 风成堆积，风积物
aeolian dune 风成沙丘
aeolian dynamics 风沙动力学
aeolian erosion 风蚀
aeolian landform 风成地貌
aeolian process 风成过程
aeolian sand 风成沙，风成砂
aeolian soil 风成土，风积土
aeolotropic 各向异性的
aeration zone 包气带
aerial camera 航空摄影机
aerial contaminant 空气污染物
aerial photograph 航片
aerial photographic gap 航摄缺漏区
aerial photography 航空摄影
aerial remote sensing 航空遥感
aerial spectrograph 航空摄谱仪
aerobic 有氧的，需氧的
aerobic decomposition 有氧分解
aerobic respiration 有氧呼吸
aerodynamic cooling 空气动力冷却
aerodynamic dissipation 空气动力耗散
aerodynamic roughness 空气动力学粗糙度
aerodynamics 空气动力学
aeroleveling 空中水准测量
aeronautical chart 航空图
aerophotography 航空摄影
aeropolygonometry 空中导线测量
aerosol 气溶胶
aerosol chemistry 气溶胶化学
aerosol composition 气溶胶成分
aerosol particle 气溶胶粒子
aerosol scattering 气溶胶散射
aerosol size distribution 气溶胶粒径分布
aerosol source 气溶胶源
aerosphere 大气层
aerostatic 空气静力的
aerostatic buoyancy 空气静力浮力
aerothermochemistry 空气热化学
aerothermodynamics 空气热力学
afforestation 造林

Agassiz Icefield 阿加西斯冰原（加拿大）
aged ridge 老化冰脊
agent of erosion 侵蚀力
agglomerated border ice 冲积岸冰
agglomeratic ice 集结冰，附聚冰
agglomerative classification 聚合分类
agglutination 黏合
aggradation 加积〔作用〕
aggradational ice 加积冰
aggregate 团聚体，聚生的
aggregation 聚合〔作用〕
agricultural culture 农耕文化
Agung〔Volcano〕 阿贡〔火山〕
air bubble 气泡
air composition 空气组分，空气成分
air diffusion 空气扩散
air freezing index 大气冻结指数
air friction 空气摩擦力
air gauge 气压计
air infiltration 空气渗透
air mass 气团
air mass trajectory 气团轨迹
air mixture 空气混合物
air monitoring 空气监测
air parcel trajectory 气块轨迹
air passage （粒雪内）气道
air permeability 透气性
air pollutant 空气污染物
air pollutant transport 空气污染物输送
air pollution 空气污染
air pollution emission 空气污染排放
air pollution episode 空气污染事件
air pollution index 空气污染指数
air quality 空气质量
air resistance 空气阻力
air specific heat 空气比热
air stratum 大气层
air temperature 气温
air thawing index 大气融化指数
airborne contaminant 空气污染物
air-borne droplet 气载液滴
airborne gravity measurement 航空重力测量
Airborne Imaging Microwave Radiometer（AIMR） 机载成像微波辐射计
airborne laser sounding 机载激光测深
airborne particulate 空气悬浮微粒
airborne radar 机载雷达
airborne sensor 机载传感器
airborne sounding 机载探空
Airborne Visible/Infrared Imaging Spectrometer（AVIRIS） 机载可见光/红外成像分光计
aircraft altimeter 航空测高计
aircraft icing 飞机结冰
Aircraft Inertial Navigation System（AINS） 飞机惯性导航系统
aircraft navigation 航空导航
Aircraft Navigation System（ANS） 飞机导航系统

aircraft sounding 飞机探空
airometer 风速计,量气计
air-plankton 大气漂浮生物
air-sea boundary process 海气边界过程
air-sea exchange 海气交换
air-sea flux 海气通量
air-sea interaction 海气相互作用
air-sea interface 海气界面
air-sea-ice interaction 气－海－冰相互作用
air-snow transfer 气－雪输送
Aitken nuclei 爱根核
Aitken nucleus counter 爱根核计尘器
Aitken particle 爱根粒子
alas 热融浅洼地,阿拉斯
Alaska Current 阿拉斯加海流
Alaska SAR Facility（ASF） 阿拉斯加雷达地面站
Alaskan high 阿拉斯加高压
Alaskan malamute 阿拉斯加雪橇狗
Alaskan Stream 阿拉斯加海流
albedo 反照率
Albic Halpi-Gelic Cambosols 漂白简育寒冻雏形土
Aleutian low 阿留申低压
Alfisol 淋溶土
Algonkian period 阿尔冈纪（元古代）
Algonkian system 阿尔冈系（元古界）
alignment survey 定线测量
alimentation （冰川）补给
alimentation area （冰川）补给区
alkali plant 耐碱植物
alkalinity 碱度,碱性
alkalinization 碱化〔作用〕
alkane 烷烃
alkene 烯,烯烃
Allegheny orogeny 阿莱干尼造山运动（晚石炭世）
Alleröd Warm Period 阿勒罗得暖期
allochthonous population 外来种群
allochthonous species 外来种
allometry method 相对生长法
allotherm 变温动物
alluvial cone 冲积锥
alluvial deposit 冲积物
alluvial facies 冲积相
alluvial fan 冲积扇
alluvial plain 冲积平原
alluvial soil 冲积土
alluviation 冲积〔作用〕
alluvion 冲积层,洪水
alluvium 冲积物,冲积层
Alluvium period 冲积期
Along Track Interferometry（ATI） 顺轨干涉
Along Track Scanning Radiometer（ATSR） 沿轨扫描辐射计
alongshore current 沿岸流
alp 冰槽肩
alpha decay α衰变
alpha ray α射线

alpine 山地的
alpine bog 山地沼泽
alpine climate 山地气候
alpine debris flows 山地泥石流,山地岩屑流
alpine desert 山地荒漠
alpine glaciation 山地冰川作用
alpine glacier 山地冰川
alpine grassland (＝rangeland) 山地草原,高寒草地
alpine mat 山地湿原,高寒湿原
alpine meadow 山地草甸,高寒草甸
alpine meadow ecosystem 山地草甸生态系统,高寒草甸生态系统
alpine meadow soil 山地草甸土,高寒草甸土
alpine periglacial environment 山地冰缘环境
alpine permafrost 山地多年冻土
alpine shrub 高寒灌丛,山地灌丛
alpine skiing 山地滑雪
Alpine Snow-Cover Analysis System (ASCAS) 高山积雪分析系统
alpine steppe 高寒草原,山地草原
alpine steppe meadow 高寒草原〔化〕草甸
alpine steppe soil 山地草原土
alpine swamp meadow 高寒沼泽〔化〕草甸
alpine terrain 山地地形
alpine tundra (＝mountain tundra) 山地冻原
alpine zone 高山带
alpinoarctic formation 北极高山群系
Alps glaciation 阿尔卑斯冰川作用
Alps mountain glacier 阿尔卑斯式山岳冰川,阿尔卑斯式山地冰川
Alps-Himalayan belt 阿尔卑斯－喜马拉雅造山带
Alps-type rock glacier 阿尔卑斯型石冰川
Altai Mountains 阿尔泰山
alternative tour 可替代性旅游,选择性旅游
altigraph 高度记录器
altimeter 高度表
altimetric measurement 测高
altimetric point 高程点
altimetry 测高法
altiplanation 高山夷平〔作用〕
altiplanation terrace 高山夷平阶地
altithermal 冰后高温期的
altitude 海拔,高程
altitude acclimation 海拔适应
altitude circle 地平纬圈
altitude gauge 高度计
altitude sick (＝mountain sick) 高山病
altitudinal distribution 垂直分布
altitudinal gradient 垂直梯度
altitudinal vicariad 垂直分布替代种

altitudinal zonality　垂直地带性
altitudinal zone　垂直带
Ambient Air Quality Standard
　（AAQS）　环境空气质量标准
Amery Ice Shelf　埃默里冰架（南极）
Amino Acid Racemization（AAR）
　氨基酸外消旋法
amino nitrogen　氨态氮
ammonification　氨化作用
ammonium ion　铵离子
ammonium nitrate　硝酸铵
ammonium sulfate　硫酸铵
amoeboid glacier　变形冰川
amphitheater terrace　圆弧形阶地
amplitude　振幅，变幅
amplitude absorption　振幅吸收
amplitude modulation　调幅
amplitude of tide　半潮差（海洋）
amplitude phase characteristic　振幅相位特性
amplitude resolution　振幅分辨率
amplitude spectrum　振幅谱
anabatic flow　上坡气流
anabatic wind　上坡风
anabolism　合成代谢
anabranch　支流
anaerobe　厌氧生物
anaerobic bacteria　厌氧细菌
anaerobic decomposition　厌氧分解
anaerobic metabolism　无氧代谢
anaerobic respiration　无氧呼吸，厌氧呼吸
anaglyphic map　互补色地图
anaglyphical stereoscopic viewing　互补色立体观察
anaglyphoscope　互补色镜
analog aerotriangulation　模拟空中三角测量
analog data　模拟数据
analog map　模拟地图
analog photogrammetry　模拟摄影测量
analog stereoplotter　模拟立体测图仪
analog waveform　模拟波形
analogue　类似物，相似体
analogue model　模拟模型
analytical aerotriangulation　解析空中三角测量
analytical orientation　解析定向
analytical photogrammetry　解析摄影测量
analytical plotter　解析测图仪
analytical precision　解析精度
analytical rectification　解析纠正
anchor ice　岸固冰
ancient glaciation　古冰川作用
ancient landform　古地形
ancient map　古地图
anelasticity　滞弹性
aneroid barograph　空盒气压表，空盒气压计
aneroid barometer　空盒气压表
angiosperm　被子植物
angular frequency　角频率
anhydrobiosis　低湿休眠

animal husbandry　畜牧业
animated mapping　动画制图
anion　阴离子
anisotropic absorption　非均质吸收
anisotropic index　各向异性指数
anisotropism　各向异性
anisotropy　各向异性
annealing algorithm　退火算法
annual course　年内过程
annual layer　年层
annual net productivity　年净生产力
annual primary productivity　年初级生产力
annual runoff　年径流量
annual species　一年生种
annual uptake　年吸收量
annuation　（生物数量）年变化
anomalous fading　（测年信号）异常衰减，异常衰退
anomaly　异常，距平
anoxia　缺氧，缺氧症
anoxic　缺氧的
Antarctic　南极的
Antarctic air mass　南极气团
Antarctic anticyclone　南极反气旋
Antarctic Bottom Water（AABW）　南极底层水
Antarctic Circle　南极圈
Antarctic Circumpolar Current（ACC）　南极绕极流
Antarctic Circumpolar Trough　南极环极槽
Antarctic Circumpolar Water　南极绕极水
Antarctic Circumpolar Wave　南极绕极波
Antarctic Cold Reversal（ACR）　南极降温逆转
Antarctic Convergence Zone（ACZ）　南极辐合带
Antarctic divergence　南极辐散〔带〕
Antarctic front　南极锋
Antarctic high　南极高压
Antarctic Ice Sheet　南极冰盖
Antarctic intermediate water　南极中层水
Antarctic Mapping Mission（AMM）　南极制图计划
Antarctic oscillation　南极涛动
Antarctic ozone depletion　南极臭氧损耗，南极臭氧洞
Antarctic ozone distribution　南极臭氧分布
Antarctic ozone hole　南极臭氧洞
Antarctic Peninsula　南极半岛
Antarctic Polar Front（APF）　南极极锋
Antarctic polar vortex　南极极涡
antarctic sagina japonica（*colobanthus quitensis*）　南极漆姑草
Antarctic sea smoke　南极海蒸汽雾
Antarctic stratospheric vortex　南极平流层涡旋
Antarctic surface water　南极表层水
Antarctic Treaty Consultative Parties（ATCPs）　南极条约协商国

antarctic usnea (＝usnea antarctica) 南极石萝
Antarctic zone 南极区
Antarctica 南极洲
antenna 天线
antenna beam 天线波束
antenna beam angle 天线波束方向角
antenna beam width 天线波束宽度
antenna directivity 天线方向性
antenna gain 天线增益
antenna height 天线高度
antenna pattern 天线方向图
antenna temperature 天线温度
Anthracolithic period 大石炭纪（石炭二叠纪）
Anthracolithic system 大石炭系（石炭二叠系）
anthropobiology 人类生物学
anthropocentrism 人类中心说，人类中心主义
Anthropocene 人类纪
anthropogenic 人为的
anthropogenic climate change 人为气候变化
anthropogenic disturbance 人为干扰
anthropogenic emission 人为排放
anthropogenic factor 人为因素
anthropogenic forcing 人为强迫
anthropogenic stress 人为胁迫
anthropogeography 人类地理学
anthroposphere 人类圈
antiboreal 南方的

anticipatory adaptation 预期适应
anticipatory policy 预期政策
anticyclogenesis 反气旋生成
anticyclolysis 反气旋消散
anticyclone 反气旋
anticyclone divergence 反气旋辐散
anticyclone inversion 反气旋逆温
anticyclone ridge 反气旋脊
anticyclone subsidence 反气旋下沉
anticyclone vortex 反气旋涡旋
anticyclone vorticity 反气旋涡度
Anticyclone Vorticity Advection (AVA) 反气旋涡度平流
anticyclonic circulation 反气旋式环流
anticyclonic gyre 反气旋式涡旋
anti-erodibility 抗蚀性
anti-freezing 抗冻
anti-ice structure 抗冰结构
anti-icer 防冰器
anti-icing 防积冰，防冰冻的
anti-icing system 防结冰系统
antineutrino 反中微子
antipleion 负偏差中心，负距平中心
aperiodic oscillaiton 非周期振动
aperiodicity 非周期性
apparent diffusion coefficient 视扩散系数
apparent heat capacity 视热容量
apparent horizon 可见地平线
apparent quantum efficiency 视量子效率
apparent temperature 视温度

apparent wave period 视波周期，可见波周期
appearance of ice 初冰
applanation 蚀平作用
Application Technology Satellite（ATS） 应用技术卫星
applied cartography 应用地图学，实用地图学
apposed glacier 合流冰川
approximate adjustment 近似平差
approximate freezing index 近似冻结指数
approximate thawing index 近似融化指数
apron 冰裙
apronlike 冰川沉积堆状的
aqua 水，溶液
Aqua satellite 阿夸卫星
aqualf 潮湿淋溶土
aquatic 水生的，水产的
aqueoglacial（＝fluvioglacial＝glacioaqueous＝glaciofluvial） 冰水的
aqueoglacial deposit 冰水沉积
aqueous 水的，水成的
aqueous deposit 水成沉积
aqueous environment 水生环境，液相环境
aquept 潮湿始成土
aquert 潮湿变性土
aquic haplorthels 含水弱育正常寒冻土
aquic haploturbels 含水弱育扰动寒冻土

aquic molliturbels 含水暗沃扰动寒冻土
aquic mollorthels 含水暗沃正常寒冻土
aquic ummrorthels 含水暗瘠正常寒冻土
aquiclude 不透水层，滞水层
aquifer 含水层，蓄水层
aquiferous 含水层的，蓄水层的
aquifuge 不透水层
aquitard 弱透水层
aquiturbels 含水扰动寒冻土
aquorthels 含水正常寒冻土
Arapahoe rock glacier 阿拉巴霍石冰川（科罗拉多）
arborescent 树枝状的
archaea 古生菌
Archaean Era（＝Archeozoic Era） 太古代
Archaean Group 太古界
archaeological dose（AD） 考古剂量
Archeozoic Era（＝Archaean Era） 太古代
archipelagic apron 列岛裙
arctalpine 北极高山的
arctalpine community 北极高山群落
arctalpine flora 北极高山植物区系
Arctic 北极
Arctic〔sea〕smoke 北极烟雾
Arctic air mass 北极气团
Arctic anticyclone 北极反气旋
Arctic Athabaskan 北极阿萨巴斯卡（印第安部落）

Arctic Athabaskan Council 北极阿萨巴斯卡委员会
Arctic blackout 北极无线电衰竭
Arctic bottom water 北极底层水
Arctic charr (*salvelinus alpinus*) 北极鲑鱼
Arctic Cinclidium Moss (*cinclidium arcticum*) 北极北灯藓
Arctic Circle 北极圈
Arctic climate 北极气候
Arctic current 北冰洋流
Arctic desert 北极荒漠
Arctic ecosystem 北极生态系统
Arctic fog 北极雾
Arctic grassland 北极草原,北极草地
Arctic grayling (*thymallus arcticus*) 北极茴鱼
Arctic haze 北极霾
Arctic heath 北极石南灌丛
Arctic high 北极高压
Arctic indigenous people 北极土著人
Arctic intermediate water 北极中层水
Arctic lichen barren 北极地衣荒地
Arctic mat grassland 北极垫状草地
Arctic meadow 北极草甸
Arctic mist 北极霭,北极冰雾
Arctic Ocean 北冰洋
Arctic Oscillation (AO) 北极涛动
Arctic plant 北极植物
Arctic polar front 北极锋
Arctic Pole 北极
Arctic shrub 北极灌丛
Arctic skuas (*stercorarius parasiticus*) 北极贼鸥
Arctic stratospheric vortex 北极平流层涡旋
Arctic surface water 北极表层水
Arctic tree line 北极树线
Arctic tropopause 北极对流层顶
Arctic tundra 北极冻原
Arctic tundra plant 北极冻原植物
Arctic tundra subzone 北极冻原亚带
Arctic warming 北极变暖
Arctic wind 北极风
Arctic wolf (*canis lupus arctos*) 北极狼
Arctic zone 北极带,北寒带
Arctic-alpine 北极高山区
Arctic-alpine dwarf shrub 北极高山矮灌丛
Arctic-alpine life zone 北极高山生物带
arcticization 北极装备,低温装备
Arcto-Tertiary flora 北极第三纪植物区系
ARDC model atmosphere 标准大气
area velocity method 面积测流法
areal differentiation 地域分异
areal oceanography 区域海洋学
areal rainfall 面雨量
arete 刃脊
argic calci-cryic aridosols 黏化钙积

寒性干旱土
argic cryi-ustic isohumosols 黏化寒性干润均腐土
argillic horizon 黏化层,粘土层
argiorthels 黏化正常寒冻土
argon 氩气
argon-39 dating 39Ar 测年
argon-40 dating 40Ar 测年
arid 干旱的
arid climate 干旱气候
arid landform 干旱地形
arid regions 干旱区
aridic epipedon 干燥表层
aridification 干旱化
aridity 干燥度
aridity factor 干燥因子
aridity index 干燥度指数
artesian aquifer 承压含水层
artesian water 承压水
artificial additional dose (AAD) 人工附加剂量
artificial freezing method 人工冻结法
artificial frozen soil 人工冻土
artificial ground freezing 人工土冻结
artificial ice stadium 人造冰场
artificial rain 人工降雨
artificial snow-making 人工造雪
ascending current 上升流
ascending motion 上升运动
asgruben 融冰洼地
asperity 粗糙度
asphalt pavement 沥青路面
aspirated psychrometer 通风干湿表
assemblage 集群,集合物
assimilation 同化〔作用〕
assimilation efficiency 同化效率
assimilation product 同化产品
assimilation rate 同化率
assumed coordinate system 假定坐标系
astrolabe 等高仪
astrometry 天体测量学
astronomical azimuth 天文方位角
astronomical constant 天文常数
astronomical coordinates 天体坐标
astronomical factor 天文因子
astronomical positioning system 天文定位系统
astronomical radiation 天文辐射
astronomical theodolite 天文经纬仪
astronomical theory （冰期）天文理论
asymmetric valley 不对称谷
asymmetry 非对称性
asynchronous 不同步的
athermobiosis 低温休眠
Atlantic subtropical dipole 大西洋副热带偶极子
atmidometer 蒸发计
atmosphere 大气,大气圈
atmosphere inverse-radiation 大气逆辐射
atmosphere-ocean interaction 海气相互作用
atmospheric aerosol 大气气溶胶
atmospheric attenuation 大气衰减

atmospheric background 大气本底
atmospheric boundary layer（ABL）大气边界层
Atmospheric Brown Cloud（ABC）大气棕色云
atmospheric chemical composition 大气化学成分
atmospheric chemistry 大气化学
atmospheric correction 大气校正
atmospheric counter radiation 大气逆辐射
atmospheric dust 大气粉尘
atmospheric environment 大气环境
atmospheric humidity 大气湿度
atmospheric ozone 大气臭氧
atmospheric path radiance 大气路径辐射
atmospheric photochemistry 大气光化学
atmospheric photolysis 大气光解
atmospheric pollutant 大气污染物
atmospheric pollution 大气污染
atmospheric profiling 大气廓线
atmospheric radiation 大气辐射
atmospheric radioactive substance 大气放射性物质
atmospheric refraction 大气折射
atmospheric structure 大气结构
atmospheric suspended matter 大气悬浮物
atmospheric trace gas 大气痕量气体
atmospheric transmissivity 大气透过率
atmospheric transport 大气传输
atmospheric turbidity 大气浑浊度
atmospheric window 大气窗
Attenberg limit 阿氏限度（黏性土的可塑性及界限含水量）
attitude parameter 姿态参数
attitude-measuring sensor 姿态测量遥感器
attribute 属性
attribute accuracy 属性精度
aufeis 冰锥
aurora australis 南极光
aurora borealis 北极光
aurora polaris 极光
auroral arc 极光弧
auroral bands 极光带
auroral cloud 极光云
auroral corona 极光冕
auroral electrojet index 极光带电急流指数
auroral excitation mechanism 极光激发机理
auroral green line 极光绿谱线
auroral hiss 极光嘶声
auroral infrasonic wave 极光次声波
auroral ionosphere 极光带电离层
auroral magnetic disturbance 极光磁扰
auroral oval 极光卵形环
auroral particle 极光粒子
auroral particle precipitation 极光粒子沉降
auroral rays 极光射线

auroral spectrum 极光光谱
auroral storm 极光暴
auroral streamer 极光幂
auroral substorm 极光亚暴
auroral zones 极光地带
austral summer 南半球夏季
automatic coordinate plotter 自动坐标测图仪
automatic triangulation 自动空中三角测量
Automatic Weather Station（AWS）自动气象站
autonomous adaptation 自适应
autotrophic respiration 自养呼吸
available moisture （植物）可吸收土壤水分
available nutrient 有效养分
avalanche 雪崩
avalanche boulder tongue 雪崩（砾）石舌

avalanche cone 雪崩锥
avalanche fan 雪崩堆积扇
avalanche glacier 雪崩冰川
avalanche moraine 雪崩碛
avalanche snow 雪崩雪
avalanche snow patches 雪崩雪斑
avalanche talus 雪崩倒石堆
avalanche track 雪崩路径
avalanche trigger zone 雪崩触发带
avalanche trough（＝avalanche chute）雪崩槽
avalanche zone 雪崩区
avalanching 雪崩，冰崩
azimuth 方位角
azimuth shift 方位偏移
azimuthal projection 方位投影
azonal vegetation 非地带性植被
azonality 非地带性
Azores High（AH）亚速尔高压

B

baby glacier 雏形冰川
Bach Ice Shelf 巴赫冰架（南极）
bacillariophyta 硅藻门
back country 边缘地区
back pressure 背压
back scattering 后向散射
back slope 后坡

backfill 回填
backfire antenna 背射天线
background air 本底空气
background atmosphere 本底大气
background concentration 本底浓度
background field 背景场
background level 本底水平

background noise　背景噪声
background pollution　本底污染
background radiation　背景辐射
background station　本底〔观测〕站
background survey　本底调查
background visibility　本底能见度
backpacker　背包旅行者
back-propagation　后向传播
backscatter　后向散射
backscattering coefficient　后向散射系数
backward scattering　逆向散射
backwearing thermokarst　溯源热熔卡斯特
bacterial ferredoxin　细菌铁氧化还原蛋白
baculate　（孢粉）棒状雕纹
badland thermokarst terrain　荒原热喀斯特地形
baffling wind　无定向微风
balance velocity　平衡速度
ballistic camera　弹道摄影机
ballistic photogrammetry　弹道摄影测量
balloon remote sensing　气球遥感
balloon-sonde　气球探空
band ratio　波段比〔值〕
band transformation　波段变换
banded berg　带状冰山
bandpass filter　带通滤波
bandpass-filtered data　带通滤波资料
band-rejection filter　带阻滤波器
bandwidth　带宽,频带宽度
bank moraine　岸碛
banked-up water　壅水
Banqiao Stage　板桥期
Baode Stage　保德期
barbed arrow　风矢（风向、风速填图符号）
barber　大风雪,冷风暴（美国和加拿大地区）
barchan ridge　新月形沙垄
barchan（＝barchane＝barkhan）新月形沙丘,新月形雪堆
bare ice　裸冰
bare ice field　裸冰原
bare ice zone　裸冰区
bare moraine　裸冰碛
bare soil　裸地
Barents-Kara Ice Sheet　巴伦支海－喀拉海冰盖
barley liquor　青稞酒
Barnes Ice Cap　巴纳斯冰帽
baroclinic effect　斜压效应
barogram　气压自记曲线
barograph　气压计
barometer　气压表
barometer column　水银柱
barometer correction　气压表订正
barometer level　气压表高度
barometric　气压（的）,气压计（的）
barometric altimeter（＝barometric altimetry）　气压高度表
barometric depression（＝barometric low）　低压

barometric equation (= barometric formula)　压高公式
barometric gradient　气压梯度
barometric height　气压高度
barometric high　高压
barometric leveling　气压高度测量
barometric low　低压
barometrograph　气压自动记录仪,气压计
barometry　气压测定法
baroreceptor　气压传感器
barosensor　气压传感器
barothermograph　气压温度计
barothermohygrograph　气压温度湿度计
barothermohygrometer　气压温度湿度表
barothermometer　压温表
barotron　气压传感器
barranca　峡谷,深谷
barren zone　裸岩区
barrens　半荒漠斑块
barrier berg　平顶冰山
barrier ice　冰障
barrier iceberg　平顶冰山
basal boundary condition　底部边界条件
basal crevasses　（冰川）底部裂缝
basal cryopeg　冰下湿寒土
basal debris　（冰川）底部岩屑
basal deformation　（冰川）底部变形
basal drag　（冰川）底部阻滞
basal freezing　（冰川）底部冻结

basal freezing rate　（冰川）底部冻结速率
basal friction　（冰川）底部摩擦
basal grinding　（冰川）基底磨蚀
basal heave　（冰川）基底式冻胀
basal ice　底冰
basal ice layer　底部冰层
basal melt rate　底部融化速率
basal melting　（冰川）底部融化
basal melting rate　底部融化速率
basal sapping　（冰川）底部掘蚀,（冰川）底部刨蚀
basal sediment　（冰川）底部沉积
basal shear stress　（冰川）底部剪应力
basal sliding　（冰川）底部滑动
basal stratigraphy　（冰川）底部冰层理
basal stream　冰下水流
basal stress　（冰川）底部应力
basal temperature　（冰川）底部温度
basal thermal gradient　底部热梯度
basal thermal regime　底部热状态
base drag　底部阻力
base flow (= base runoff)　基流
base line　基线〔准〕
base line network　基线网
base moraine hill　底碛丘陵
base of permafrost　多年冻土基底,多年冻土底板
base period　基流期
base pressure characteristic　底部

压力特征
base pressure coefficient 底部压力系数
base pressure distribution 底部压力分布
base-height ratio 基一高比
baseline estimation 基线估计
baseline measurement 基线〔准〕测量
baseline photographic 摄影基线
baseline radar 基线雷达
baseline scenario 本底情景
baseline station 基准站
basement contour 基底等深线
basic gravimetric point 基本重力点
basin adaptability 流域适应能力
basin divide 分水岭
basin length 流域长度
basin outlet 流域出口
basin recharge 流域补给
basin time lag 流域滞时
bataclave 防寒帽
bathymetric chart 水深图
bathymetric contour 等深线
bathymetric measurement 水深测量
bathymetric survey 水深调查
bathymetry 水深测量仪,水深测量法
batbathythermograph 深温仪
bay ice 海湾冰
beach ice 海滩冰
beaded lake 串珠状湖
bearing capacity 承载力

Beaufort force 蒲福风力
bed deformation （冰川)底部变形
bed elevation 冰床高程
bed roughness 冰床粗糙度
bed separation index 底床分离指数
bed structure 冰床结构
bed topography 冰床地形
bedded structure 层状构造
bedding 层理作用
bedding surface 层面
bedrock 基岩
bedrock adjustment 基岩调整
bedrock depression 基岩凹陷
bedrock reflection 基岩反射
bedrock roughness 基岩粗糙度
bedrock topography 基岩形态
Beekley gauge 贝克来雨量计
Beer's law 比尔定律
Before Christ（B.C.) 公元前
belayer 攀登保护者
bellicatter 冰湖,冰栅
Ben religion 苯教（宗教)
benchmark 基准
bending failure 弯曲破坏
bending ice force 弯曲冰力
benthic 底栖生物
benthithermoprobe 海底温度仪
benthonic 海底的,底栖的
benzene 苯
berg bit 小冰山
bergschrund〔e〕 冰川后壁裂隙
berg-till 浮冰碛
bergy bit 浮冰块（尺寸小于5 m)

bergy seltzer （冰川冰在水中融解时气泡发出的）咝咝声
Berkner Island 伯克纳岛（南极洲）
berm snow 雪护道
berm soil 土护道
Beryllium-10 ^{10}Be
Beryllium-10 dating ^{10}Be 测年
beset （海冰）围困
Bessel ellipsoid 贝塞尔椭球
Beta (β) dacay β 衰变
Beta (β) particle β 粒子
Beta (β) ray β 射线
Beta (β) activity β 活化度
Bhaskara 巴斯卡拉卫星
bias correction 偏差订正
bicarbonate 碳酸氢盐
bi-cubic convolution method 双三次卷积法
bidirectional albedo 双向反照率
bidirectional reflectance 双向反射
bidirectional reflectance distribion function (BRDF) 双向反射率分布函数
bidirectional reflectance factor 双向反射因子
biennial ice 二年冰
bifurcation point 分叉点
big leaf model 大叶模型
big-scale-centralization 大尺度集聚
biocenosis 生物群落
biochemical oxygen demand (BOD) 生化需氧量
biochemistry 生物化学
biochore 植物区气候（柯本气候分类）
biochronology 生物年代学
bioclimate 生物气候
bioclimate zonation 生物气候分区
bioclimatograph 生物气候图
bioclimatology 生物气候学
bioclock 生物钟
biocompatibility 生物适应性
biocycle 生物循环
biodiversity 生物多样性
bioecology 生物生态学
biofeedback 生物反馈
biogenic deposit 生源沉积
biogenic ice nucleus 生源冰核
biogenic ooze 生源软泥
biogenic trace gases 生源痕量气体
biogeochemical cycle 生物地球化学循环
biogeochemistry 生物地球化学
biogeographic region 生物地理区域
biogeography 生物地理学
biogeosphere 生物地理圈
bioinvasion 生物入侵
biolimiting element 生物限制性元素
biological dating method 生物学定年法
biological feedback 生物反馈
biological production 生物生产〔量〕
biological pump 生物泵
biological weathering 生物风化
biomass 生物量〔质〕

biomass accumulation 生物量累积
biomass burning 生物质燃烧
bioorganic phosphorus 生物有机磷
biophenology 生物物候学
biophysical chemistry 生物物理化学
biophysics 生物物理学
biosphere 生物圈
biostratigraphy 生物地层学
biota 生物群落
biotic district 生物地理区
biotic ecotype 生物生态型
biotic factor 生物因子
biotic province 生物地理省
biotope 群落生境
bipolar distribution 双极分布
biscuit board topography （冰川）蚀余地形
bistatic radar 双基雷达
bistatic Synthetic Aperture Radar (SAR) 双基合成孔径雷达
bit 钻头
bivouac tent 轻便帐篷
black body radiation 黑体辐射
black box model 黑箱模型
black box theory 黑箱理论
black carbon 黑碳
black carbon aerosol 黑碳气溶胶
black frost 黑霜
black ice 雨凇,黑冰
black-and-white infrared image 黑白红外影像
black-and-white photography 黑白摄影

blind drainage area 闭流区
blind lead （冰间）死水道
blind valley 盲谷
blizzard 暴风雪
blizzard fatality 暴风雪伤亡
blizzard warning 暴风雪预警
block adjustment 区域网平差
block diagram 块状图
block field (＝block sea) 石海
block slide 整体滑动
block slope 石流坡
block stream 石河
block stripe 石条
Block-5D 布洛克－5D卫星
block-schollen movement 冰块运动
blood rain 血雨
blood snow 血雪
blowhole 风蚀穴
blowing dust 扬尘
blowing snow 风吹雪,高吹雪 （2 m 高度以上）
blowout 风蚀凹地
blowout pit 风蚀坑
blue haze 蓝霾
blue ice 蓝冰
blue-ice area 蓝冰带
bob-run 雪橇滑道
bobsledder 有舵雪橇运动员
bobsledding 有舵雪橇运动
bobsleigh chute 有舵雪橇滑道
bobsleigh (＝bobsled) 长(大)雪橇
body temperature regulation 体温调节

bog soil 沼泽土
bogginess 沼泽性
bogging 沼泽土化
boggy 沼泽的
Bogoslof Volcanic Island 博戈斯洛夫火山岛（阿拉斯加）
bogs 苔藓泥炭沼泽
bogus data 假想资料
boiling spring 沸泉
boit 胀式岩石锥
Bolling-Allerod Warm Period（BAW）博令－阿洛德暖期
Boltzmann constant 波尔兹曼常数
bond strength 结合强度
border ice 岸冰
border moraine 边碛
bore bit 钻头
bore core 岩芯
Boreal 北方气候期
boreal climate 北方气候
boreal climatic phase 北方气候期
boreal ecosystem 北方生态系统
boreal forest（＝boreal Taiga） 北方森林（北方泰加林）
boreal kingdom 北方界
boreal period 北方期
boreal pole 北极
boreal region 北方区
boreal summer 北半球复季
boreal wetlands 北方湿地
boreal winter 北半球冬季
boreal woodland 北方林地
boreal zone 北方区

boreas 北风
borehole 钻孔，井孔
borehole camera 钻孔摄像机
borehole deflectometer 钻孔挠度计
borehole deformationmeter 钻孔形变计
borehole dilatometer 钻孔膨胀计
borehole expansion probe 钻孔膨胀探测仪
borehole extensometer 钻孔张力仪
borehole inclinometer 钻孔倾斜仪
borehole logging 钻孔记录
borehole penetrometer 钻孔贯入仪
borehole stressmeter 钻孔应力计
borehole temperature 钻孔温度
borehole temperature profile 钻孔温度剖面
boring 钻探，打钻
boring log 钻孔记录，钻孔柱状图
boring rig 钻探机具
bottom conduit 底部水道
bottom current 底层流
bottom current meter 底层流速仪
bottom deposit 底部沉积
bottom discharge 底部排水
bottom friction 底部摩擦
bottom ice 底冰
bottom moraine（＝ground moraine）底碛
bottom pressure 底部压力
bottom structure 底部结构
bottom temperature 底部温度
bottom topography 底部地形

bottom tractive force 底部拖力
bottom view 底视图
bottom water (大洋)底层水
bottom-dwelling 底栖的
bottomset bed 底积层
bottom-up model 自下而上模式
boudary element method 边界元法
boulder 漂砾,巨砾
boulder bed 巨砾层
boulder channel 漂砾河道
boulder clay (冰川)泥砾
boulder fan 冰砾扇
boulder pavement 砾石地
boulder-field 巨砾原,砾石原
boulder-wall 冰砾壁
bouldery mantle 巨砾覆盖层
bound water 束缚水
boundary current 边界流
boundary layer 边界层
boundary layer climate 边界层气候
bowl-shaped frost table 碗状冻结面
brackish ice 咸水冰
brash 破碎的
brash ice 碎冰
brave west winds 咆哮西风
breaking point 断裂点,强度极限
breaking strain 破坏应变
breaking strength 破坏强度
breaking stress 破坏应力
breakup 解冻,开河
breakup date 解冻日
breakup forecast 解冻预报
breakup period 解冻期
breakup season 解冻季节
brightness 亮度
brightness temperature 亮温
brine lake 卤水湖
brine rejection 析盐
brine salinity 卤水盐度
brittle fracture 脆性断裂
brodelboden 冻融包裹土,冰卷泥
broiling 酷热天
broken belt 碎冰带
broken layer 断裂层
brown snow 棕色雪
bryochore 苔原区,冻原区,有生物区
bryophyte 苔藓植物
bubble (冰内)气泡
bubble formation 气泡形成
bubble growth 气泡发育
bubble radius 气泡半径
bubble-free 无气泡的(冰)
bubble-rich ice 富含气泡冰
bubbly ice (含气)泡冰
bucket temperature (海洋)表面水温
bucket thermometer (海洋)表面水温表
built platform 堆积台地
built terrace 堆积阶地
bulb glacier 扇状冰川,宽尾冰川
bulge hypothesis (冰盖)扩张假说
bulk compressibility 体积压缩性
bulk density 体积密度,容重
bulk elasticity 体积弹性

bulk strain 体积应变
bummock 倒置冰丘
buoy 浮标
buoyancy velocity 浮力速度
burial dating 埋藏测年,埋藏定年
buric fibri-permagelic histosols 埋藏纤维多年冻结有机土
buried channel 埋藏河道
buried glacier ice 埋藏冰川冰
buried ice 埋藏冰
buried organic layer 埋藏有机质层
buried ridge 埋藏山脊
buried river 埋藏河
buried soil 埋藏土壤
buried soil horizon 埋藏土层
buried terrace 埋藏阶地
buttered tea 酥油茶

C

caclcic halpi-gelic cambosols 钙积简育寒冻雏形土
caclcic molli-gelic cambosols 钙积暗沃寒冻雏形土
cadder 冰水壶穴
calcaric geli-orthic primosols 石灰寒冻正常新成土
calcaric geli-sandic primosols 石灰寒冻砂质新成土
calcaric halpi-gelic cambosols 石灰简育寒冻雏形土
calcaric matti-gelic cambosols 石灰草毡寒冻雏形土
calcaric molli-gelic cambosols 石灰暗沃寒冻雏形土
calcic anhyorthels 钙积脱水正常寒冻土
calcic cryi-ustic isohumosols 钙积寒性干润均腐土
calcic matti-gelic cambosols 钙积草毡寒冻雏形土
calm belt 无风带(气象)
calve (冰)崩解,分离
calved ice 崩塌冰
calving 冰崩
calving front 崩解面
calving glacier 崩解冰川
calving rate 崩解速率
can ice 大块冰
candle ice(＝penknife ice) 烛冰,小刀状冰,冰指
canopy 冠层
canyon wind 谷风
capillary water 毛细水
captive yaks(*poephagus grunniens*) 牦牛

carbohydrate 碳水化合物
carbon assimilation 碳同化
carbon balance 碳平衡
carbon budget 碳收支
carbon cycle 碳循环
carbon cycle model 碳循环模型
carbon dating 碳定年法
carbon dioxide cycle 二氧化碳循环
carbon dioxide equivalence（CDE） 二氧化碳当量
carbon dioxide fertilization 二氧化碳施肥
carbon dioxide fixation 二氧化碳固定
carbon dioxide sensitivity 二氧化碳敏感性
carbon disulfide 二硫化碳
carbon emission 碳排放〔量〕
carbon fixation 固碳
carbon flux 碳通量
carbon isotope 碳同位素
carbon isotope ratio 碳同位素比率
carbon loss 碳丢失
carbon monoxide 一氧化碳
carbon neutral 碳中和
carbon nitrogen ratio 碳氮比
carbon nutrition 碳营养
carbon pool 碳库
carbon sequestration 碳封存
carbon sink 碳汇
carbon source 碳源
carbon stock（＝storage） 碳储存
carbon tetrachloride 四氯化碳

carbon-14 dating ^{14}C 测年，碳-14测年
carbon-14-labeled tracer ^{14}C 标记示踪物
carbonate 碳酸盐
carboxylation 羧〔基〕化，羧化作用
cartography 地图学
cascade stairway 冰阶坎
cascading glacier 瀑布冰川
catastrophe 灾变
catastrophic event 灾难性事件
catch drain 集水沟
catch pit 集水坑
catch ratio 捕捉率
catchment 流域，集水区
catchment area survey 汇水面积测量
catchment glacier（＝snowdrift glacier） 吹雪冰川，洼地冰川
cation 阳离子
cave deposit 洞穴沉积物
cave ice 洞穴冰
cavity （冰内、冰下）空腔
cell of brine 盐水泡（冰中）
Cenozoic Ice Age 新生代冰期
channel capacity 河道容量
channel control 河道断面
channel roughness 河道粗糙度
channel slope 河道比降
charge-coupled device（CCD） 电荷耦合器件
chatter marks 冰川擦痕
chemical geography 化学地理学
chemical oxygen demand（COD）

化学需氧量
chemical weathering　化学风化
chilling injury　冷害
China Grassland Transect（CGT）
　　中国草地样带
China-Brazil Earth Resources Satellite（CBERS）　中巴地球资源卫星
Chinese caterpillar fungus（*ophiocordyceps sinensis*）　冬虫夏草
Chinese Geodetic Stars Catalogue（CGSC）　中国大地测量星表
chionophile　嗜雪植物,喜雪植物
chionophobous plant　厌雪植物
Chlorine-36 dating　^{36}Cl 测年
Chlorofluorocarbons（CFCs）　氟利昂,氯氟碳化合物
Chlorofluoromethane（CFM）　氯氟甲烷
chondrite　球粒陨石
chronostratigraphy　年代地层学
cinderic geli-orthic primosols　火山渣寒冻正常新成土
circular snow patch　圆形雪斑
circumpolar circulation　绕极流
circumpolar cyclone　绕极气旋
circumpolar distribution　环极分布
circumpolar westerlies　环极西风带
cirque　冰斗
cirque cutting　冰斗切刻
cirque erosion　冰斗侵蚀
cirque floor　冰斗底
cirque glacier　冰斗冰川

cirque lake　冰斗湖
cirque moraine　冰斗冰碛
cirque platform　冰斗平台
cirque stairway　冰斗阶梯
cirque step（＝cirque terrace）　冰斗阶地
cirque-valley glacier　冰斗山谷冰川
Clean Air Act Amendments（CAAA）　空气清洁法修正案
Clean Development Mechanism（CDM）　清洁发展机制
clear air turbulence（CAT）　晴空湍流
clear ice　洁净冰,透明冰
clear sky precipitation　晴空降水
climate abnormality　气候异常性
climate analog　气候类比
climate archive　气候档案,气候记录
climate change　气候变化
climate change commitment　气候变化承诺
climate evolution　气候演化
climate fluctuation　气候波动
climate indicator　气候指示器
climate model-based scenarios　基于气候模式的情景
climate periodicity　气候周期性
climate prediction　气候预测
climate projection　气候预估
climate proxy　气候代用指标
climate reconstruction　气候重建
climate resources　气候资源
climate sensitivity　气候敏感性

climate snow line 气候雪线	climatography 气候志
climate stability 气候稳定性	climatology 气候学,气候态
climate system 气候系统	climatostratigraphy 气候地层学
climate threshold 气候阈值	climax community 顶极群落
climate tourism resources 气候旅游资源	climax forest 顶极森林
climate vacillation 气候脉动	climax formation 顶极群系
climate variability 气候变率	climbing boots 登山靴
climate variation 气候变动	climbing rope 登山绳
climate warming 气候变暖	climbing shoes 爬岩鞋
climate adaptation 气候适应	climbing trousers 登山裤
climatic anomaly 气候异常	climotope 气候生境
climatic climax vegetation 气候顶极植被	closed lake 内陆湖,封闭湖
climatic element 气候要素	closed pack ice 密集浮冰
climatic factor 气候因子	closed system 封闭系统
climatic geomorphology 气候地貌学	closed system pingo 封闭型冰丘
climatic gradient 气候梯度	closed talik 非贯穿融区
climatic impact 气候影响	closed-cavity ice 封闭洞穴冰
climatic instability 气候不稳定性	closed-system freezing 封闭系统冻结
climatic optimum 气候适宜〔期〕,气候最适期	close-path eddy covariance (CPEC) 闭路涡度相关
climatic potential productivity 气候生产潜力	cloud albedo 云反照率
climatic productivity 气候生产力	cloud condensation nuclei (CCN) 云凝结核
climatic record 气候记录	cloud cover 云覆盖
climatic sensitivity 气候敏感性	cloud feedback 云反馈
climatic zonality 气候地带性	cloud radiative forcing 云辐射强迫
climatic zone 气候带	cloudburst 暴雨,大雨
climatization 适应气候	cloudburst flood 暴雨洪水
climatoecological type 气候生态型	coarse grain 粗颗粒
climatograph 气候图	coarse-grained firn 粗粒雪
	coastal comb 海岸冰堆(抛到岸上的海冰)

coastal ice 沿岸冰,海岸冰
coastal pressure ridge 沿岸冰脊
coherent radar 相干雷达
cohesive force 内聚力
cold advection 冷平流
cold air 冷空气
cold air mass 冷气团
cold anticyclone 冷性反气旋
cold bath 冷浴
cold cap 冷冠
cold conveyor belt 冷空气输送带
cold crack temperature 冷裂温度
cold cyclone 冷性气旋
cold damage 冷灾
cold deformation 冷变形
cold desert 冷荒漠,寒带荒漠
cold drain 冷空气下沉
cold endurance 耐寒力
cold event 冷事件
cold fog 冷雾
cold front 冷锋
cold front rain 冷锋雨
cold front thunderstorm 冷锋雷暴
cold glacier 冷型冰川
cold high 冷高压
cold injury 冷害
cold island 冷岛
Cold Land Processes Experiment (CLPX) 寒区陆面过程试验
cold low 冷低压
cold percolation zone 冷渗浸带
cold period 冷期
cold permafrost 低温多年冻土
cold phase 冷时段
cold pole 冷极
cold pool 冷池
cold regions environment 寒区环境
cold regions ecology 寒区生态学
cold regions ecosystem 寒区生态系统
cold regions engineering 寒区工程
cold regions hydrogeology 寒区水文地质学
cold regions hydrology 寒区水文学
cold resistance 抗寒性
cold season 冷季
cold soak （设备的）低温适应性
cold source 冷源
cold trough 冷槽
cold vortex 冷涡
cold wave 寒潮
cold wedge 冷楔
cold wind 冷风
cold zone 寒带
cold-air injection 冷空气侵入
cold-air outbreak 冷空气爆发,寒潮爆发
cold-based glacier 冷底冰川
collapse scar 融塌遗迹
collapsed pingo 坍塌的多年生冻胀丘,冰皋
colluvial deposit 崩积物
colluvial soil 崩积土
colluvium 崩积层
columnar ice 柱状冰

combined adjustment 联合平差
commitment period （温室气体减排的）承诺期
committed climate change 持续性气候变化（温室气体浓度稳定后的气候变化）
Common Agricultural Policy（CPA）（欧共体）共同农业政策
common-midpoint（CMP） 共中心点（测绘）
community 社区,群落
community ecology 群落生态学
community evolution 群落演化
community gross production 群落总生产量
community organization 群落组织
community sample 社区样本
community scale 社区尺度
Compact High Resolution Imaging Spectrometer（CHRIS） 紧密型高分辨率成像光谱仪
compact ice 密集冰,固结冰（海冰）
compact snow 密实雪
compacted ice edge 密集海冰外缘
compacted snow parking （修建的）停车场 压实雪
compacted snow road 压实雪（筑成的）路
compaction 压实
compactive viscosity 压缩黏滞度
compass 指南针
complete freeze-up 完全冻结（河流、湖泊中的水体全部冻结）
complete freezing 完全封冻,封冻
complete ice coverage 封冻,冰封
complex dielectric constant 复介电常数
component 成分
composite analysis 合成分析
composite ice wedge 复合冰楔
composite valley glacier 复式山谷冰川
compound cirques 复合冰斗
compound crystal 复晶
compound pancake ice 复合盘状冰（海冰）
compound valley glacier 复合山谷冰川
compressing flow 压缩流
compression strain rate 压缩应变率
compressive flow 可压缩流
compressive strength 抗压强度
compressive stress 压应力
computing grid 计算网格
concentration 浓度
concentration time 汇流时间
concrete pavement 水泥路面
concrete snow 固结雪
concurrent glaciation 同期冰川作用
condensability 凝结性
condensate 冷凝物
condensated water 冷凝水
condensation 凝结〔作用〕
condensation efficiency 凝结效率
condensation function 凝结函数

condensation heat 冷凝热,凝结热
condensation hygrometer 冷凝湿度表
condensation level (CL) 凝结高度
condensation nucleus 凝结核
condensation particle 凝结粒子
condensation process 凝结过程
condensation recharge 凝结水补给
condensation temperature 凝结温度
condensation trail 飞行凝结尾迹
condensation water 凝结水
conductivity 传导率
conduit (冰川内)管道
configuration 构造,结构
confined aquifer 承压含水层
confined flow 承压流
confined water 承压水
confined water basin 承压水盆地
confining boundary 隔水边界
confluence 汇流
confluence step 冰川角阶
confluent glacier 汇合冰川
conformal projection 等角投影
congealed depth hoar 聚合深霜
congealing point 冻结点
congelation 冻结〔作用〕
congelation ice 冻结冰
congelifract 寒冻风化
congelifractate 融冻(崩解)岩块
congelifraction 冰冻风化,融冻(崩解)作用
congeliturbate 融冻堆积物,融冻扰动土

congeliturbation 融冻泥流作用
conglomerate 砾岩
conglometated ice (叠加的)浮冰块
conglometated packing ice 凝聚浮冰(群)
conic projection 圆锥投影
conifer forest (＝coniferous forest) 针叶林
coniferous forest (＝conifer forest) 针叶林
conifruticeta 针叶灌木林
coniscope 计尘仪
conjugate points 共轭点
connecting traverse 附合导线
connection triangle method 大地三角联测
conophorium 针叶林群落
conophorophyta 针叶林植物
conregional 区域特征种的
conservation 保护
conservation measures 保护措施
conservative grazing 合理放牧
consolidated ice 叠加冰(河、湖、海冰)
consolidated layer 叠加海冰层,(冻土)固结层
consolidated packing ice 固结浮冰
consolidated ridge 叠加冰脊
consolidation 固结作用
constant wind 稳定风
Constellation of Small Satellites for Mediterranean Basin Observation (COSMO-SkyMed) 地中

海区域小卫星群观测
constituent tide 分潮
constitutive equation 本构方程
constructed wetland 人工湿地
constructional plain coast 堆积平原海岸
constructive species 建群种
consumer 消费者
contact anemometer 电接〔式〕风速表
contaminant 污染物
contamination 污染
continental aerosol 大陆性气溶胶
continental air mass 大陆性气团
continental anticyclone 大陆性反气旋
continental apron 大陆裙
continental borderland 大陆边缘地
continental climate 大陆性气候
continental drift 大陆漂移
continental glaciation 大陆性（型）冰川作用
continental glacier 大陆性（型）冰川
continental high 大陆高压
continental periglacial climate 大陆性（型）冰缘气候
continental periglacial region 大陆性冰缘区
continental scale 大陆尺度
continental shelf 大陆架
continental shelf sediment 陆架沉积物

continental slope facies 大陆坡相
continental wind 大陆风
continentality 大陆度
continuous grazing 常年放牧
continuous internal reflector （冰）连续内部反射体
continuous permafrost 连续多年冻土
continuous snow cover duration （CSCD） 连续积雪期
contour 等值线，等高线
contourite 等积岩
contraction cracking 收缩破裂
contrast enhancement 对比度增强
contrast stretching 对比度拉伸
control point 控制点
control survey 控制测量
convection 对流
convective boundary layer 对流边界层
convective condensation level （CCL） 对流凝结高度
convective mixing 对流性混合
convective precipitation 对流性降水
convective transfer 对流传输
conventional gas constant 通用气体常数
Conventional International Origin （CIO） 国际协议原点
convergence 辐合
convergent adaptation 趋同适应
convergent ice flow 辐合冰流

convergent oscillation 趋同波动
converging flow 收敛流,收敛流动
converging ice-stream ponded lake 冰汇堰塞湖
convolution 卷积
cooking snow 湿雪
cool damage 冷害
cool injury 冻伤
coolant 冷却剂
cooling 冷却作用
cooling ducts 制冷管
cooling index 制冷指数
cooling process 冷却过程
cooling rod 冷却杆(露点仪用)
cooling surface 冷却面
cooling temperature 冷却温度
cooling-power anemometer 冷却率风速表
Cooperated Enhanced Observing Period(CEOP) 全球协调加强观测
coordinate grid 坐标格网
coordinate measuring instrument 坐标量测仪
coordinate rotation 坐标轴旋转
Coordinated Universal Time(UTC) 协调世界时
coplanarity equation 共面方程
coppice stage 萌生林阶段
coppice-land 萌生林地
coprophyte 粪生植物
coprozoon 粪生动物
copse regeneration 萌芽更新

coral 珊瑚
coral reef 珊瑚礁
coregistration 配准
coreless winter 平缓冬季(指南极冬季日平均气温变化不大)
Coriolis force 科里奥利力,科氏力
corn snow (春天)玉米粒状雪
corner reflector 角形反射器
corona 日冕
Corona satellite (美国20世纪60年代开发的)成像侦察卫星
coronal mass ejection(CME) 日冕物质抛射
corpuscle 微粒
corrasion 刻蚀,磨蚀
correction 修正
correction factor 修正因子
correlated variable 相关变量
correlation analysis 相关分析
correlation coefficient 相关系数
correlation matrix 相关矩阵
correlation test 相关性检验
corrie 山凹,冰坑,冰斗
corrie(=cirque)glacier 冰斗冰川
corrosion 溶蚀,腐蚀
cosmic abundance 宇宙线丰度
cosmic dust 宇宙尘埃
cosmic microwave background radiation(CMBR) 宇宙微波背景辐射
cosmic ray 宇宙线
cosmochronology 宇宙年代学
cosmogenic 宇宙(线)成因的

cosmogenic nuclide dating 宇宙成因核素测年
cosmogenic radioisotope 宇宙成因的放射性同位素
cospectrum 协谱
cost-benefit analysis（CBA） 成本－效益分析
Coulomb friction 库仑摩擦
Coulomb slip 库仑滑动
coulometric oxidant analyzer 电量法氧化剂分析仪
counter-adaptation 逆适应
coupled system 耦合系统
coupling 耦合
coupling coefficient 耦合系数
covariance 协方差
covariance function 协方差函数
crackle 裂纹
crag and tail 鼻尾山
crater 火山口
crater cirque 火山口冰斗
cream ice 初凝冰
creep 蠕变
creep consolidation 蠕变固结
creep flow 蠕变流
creep fracture 蠕变断裂
creep function 蠕变函数
creep instability 蠕变不稳定性
creep movement 蠕动
creep rate 蠕变速率
creeping mass 蠕动块体,滑坍块体
crescent dune 新月形沙丘
crest 山脊

crevasse deposit 裂隙沉积物
crevasse depth 裂隙深度
crevasse filling 裂隙冰水沉积
crevasse hoar （冰）裂隙霜
crevasse pattern 裂隙型式
crevasse trace 裂隙痕迹
crevasses 冰裂隙
criterion 标准
critical angle 临界角
critical condition 临界条件
critical depth 临界水深
critical discharge 临界流量
critical embankment height 临界路堤高度
critical factor 关键因素
critical flow 临界流
critical speed 临界速度
critical temperature 临界温度
critical value 临界值
critical velocity 临界速度
critical zone 临界带,脆弱区
Cro Magnon Man 克罗马农人
cross polarization 交叉极化
cross section 横断面
cross subsidies 交叉补贴
cross-country skiing 越野滑雪
crossing point analysis 交叉点分析
crossover analysis 交叉分析
crossover discrepancy 交叉点偏差
cross-sectional flow 断面流量
cross-track interferometry 交轨干涉
cross-track stereo 跨轨像对（航向重叠）

crumb structure 团粒结构
crushed ice 挤压冰,碎冰
crushed rock embankment 块石路基
crushed rock-based embankment 块石基底路基
crushed rock-covered embankment 块石护坡路基
crushing failure 挤压破坏
crushing ice force 挤压冰力
crushing strength 挤压强度
crust deformation measurement 地壳形变观测
crustal movement 地壳运动
crustal stress 地壳应力
crustal tension 地壳张力
crustic epipedons 结皮表层
crust-like cryostructure 壳状冷生构造
crustose lichen 壳状地衣
cryic andosols 寒性火山灰土
cryic aridosols 寒性干旱土
crymad 极地植物
cryoalgae 冰雪藻类
cryobiochemistry 低温生物化学
cryobiology 低温生物学
cryobiosis 低温生态
cryochore 冰雪区,冰雪气候带
cryoconite 冰穴尘
cryoconite hole 单冰穴
cryogenesis 冷生作用
cryogenic 低温学的,低温冷冻
cryogenic fabric 冷生微结构
cryogenic hygrometer 低温湿度计
cryogenic lake 热融湖塘
cryogenic material 寒冻物质
cryogenic period 冷期
cryogenic process 冷生过程
cryogenic weathering 寒冻风化
cryokarst 冰冻喀斯特
cryolithology 冷岩学
cryolithosphere 冷生岩石圈
cryolithozone 冷生岩石带
cryology 冰冻学
cryopediment 寒冻麓原
cryopedology 冻土学
cryopedometer 冻土深度探测器,冻土器
cryopeg 湿寒土
cryophyte 冰雪植物
cryoplanation 寒冻夷平作用
cryoplanation terrace 寒冻夷平阶地
cryopump 低温抽气泵
cryosol 冷生土壤,寒冻土
cryosolic order 寒冻土纲
cryosphere 冰冻圈
cryosphere components 冰冻圈组成
cryosphere fluctuation 冰冻圈波动
cryospheric sciences 冰冻圈科学
cryostatic pressure 冷生静压
cryostatic process 静态冷生过程
cryostructure 冷生构造
cryosuction 冻结抽吸
cryotexture 冷生组构
cryotic ground 负温土,冷生土
cryoturbation 融冻扰动作用
cryoturbation structure 冻裂搅动

crystal fabric orientation（COF） 冰晶（主轴）取向
crystal morphology 结晶形态
crystallization 结晶作用
crystallization differentiation 结晶分异作用
crystophene （泉水上升或扩散过程形成的）地下冰
cultural heritage 文化遗产
cultural identity 文化认同
culvert icing 涵洞冰锥
cumulative ablation 累积消融
cumulative effect 累积效应
cumulative mass balance 累积物质平衡
cumulative net balance 累积净物质平衡
cumulative positive temperature 正积温
cumulative thermogradient 累积热梯度
cumulic molliturbels 堆积暗沃扰动寒冻土
cumulic mollorthels 堆积暗沃正常寒冻土
cumulic umbriturbels 堆积暗瘠扰动寒冻土
cumulic ummroethels 堆积暗瘠正常寒冻土
cup crystal 杯状冰晶
cup-generator anemometer 磁感风杯风速表
curling 冰壶（体育）
current drag force 流拖曳力
current meter 流速表，海流计
current surveying 测流
current velocity measurement 流速测量
current year's runoff 当年径流
current-meter discharge 流速仪测流
curtain aurora 帘状极光，极光帘
curvity 曲率
cwm（＝cirque） 冰斗
cyclohexane 环己烷
cyclone 气旋

D

daily extreme 日极值
daily maximum temperature 日最高温度
daily mean temperature 日平均温度
daily minimum temperature 日最低温度

daily precipitation 日降水量
daily range 日较差
daily variation 日变化
Dali Glaciation 大理冰期
dam cooling systems 大坝冷却系统
damaged ecosystem 受损生态系统
dam-break flood 溃坝洪水
dammed lake 堰塞湖
damp air 湿空气
dampness 湿度,含水量
Dansgaard-Oeschger（D-O）oscillations D-O 振荡
Dansgaard-Oeschger（D-O）cycle D-O 旋迴
Dansgaard-Oeschger（D-O）event D-O 事件
Darcian flow 达西流
Darcy's Law 达西定律
dark aqui-gelic cambosols 暗色潮湿寒冻雏形土
dark argi-cryic aridosols 暗色黏化寒性干旱土
dark calci-cryic aridosols 暗色钙积寒性干旱土
dark hapli-cryic andosols 暗色简育寒性火山灰土
dark hapli-cryic aridosols 暗色简育寒性干旱土
dark nilas 暗尼罗冰
dark reaction 暗反应
dark respiration 暗呼吸
dasymetric map 分区密度地图
data acquisition 数据采集

data acquisition instrument 数据采集仪
data acquisition system 数据采集系统
data assimilation 数据同化,资料同化
data automatically transmission 数据自动传输
data bank 资料库,数据库
data base 数据库
data capture 数据采集
data classification 数据分类
data collection 数据收集
data consolidation 数据整合
data directory 数据目录
data exchange system 数据交换系统
data extraction 资料提取
data format 数据格式
data fusion 数据融合
data grid 数据网格
data integration 数据集成
data integrity 数据完整性
data logger 数据记录仪
data mining 数据挖掘
data model 数据模型
data platform 数据平台
data processing 数据处理,信息加工
data quality control 数据质量控制
data reconstruction 资料重建
data rejection 数据剔除
data restoration 数据恢复
data sample 数据样本
data acquisition system（DAS） 数

据采集系统
data selection 数据选择
data service 数据服务
data set 数据集
data sharing 数据共享
data source 数据源
data structure 数据结构
data transfer 数据转换
data transmission 数据传输
data transport protocol (DTP) 数据传输协议
data visualization 数据可视化
data-model fusion system 数据模型融合系统
dataset catalog 数据集目录
dataset directory 数据集目录
date line 日界线,日期变更线
date of break-up 解冻日期
date of freeze-up 封冻日期
dateability 可定年性
dating 定年
dating horizon 标志年层
dating method 定年法
dating technique 测年技术
datum plane 基准面
datum state 基准态
daughter isotope (放射性元素衰变后的)子元素
dawnside auroral oval 晨侧极光卵
day temperature 白天温度,日温
day without frost 无霜日
dayside aurora 昼侧极光
dayside auroral oval 昼侧极光卵

dayside magnetosphere 昼侧磁层
day-to-day change 日际变化,逐日变化
DC resistivity sounding 直流电阻率〔法〕
de Saint-venant equations 圣维南方程组
dead glacier 死冰川,停滞冰川
dead ice 死冰
dead-ice area 死冰区
dead-ice moraine 死冰碛
deaired water 脱气水
debacle 解冻,崩溃
debris 岩屑
debris cone 表碛锥,岩屑锥,倒石锥
debris flow deposit 泥石流堆积
debris flows 泥石流
debris island 岩屑岛
debris mantled slopes 岩屑坡
debris slide 岩屑滑坡
debris slope 岩屑坡
debris tails 岩屑尾状体
debris terrace 岩屑阶地
debris-covered glacier 表碛覆盖冰川
decarboxylase 脱羧酶
decarboxylation 脱羧作用
decay 腐解〔作用〕
decay curve 衰减曲线
deciduous broadleaf forest (DBF) 落叶阔叶林
deciduous needleleaf forest (DNF) 落叶针叶林

decision support system (DSS)　决策支持系统
deckenschotter　冰水沉积平原
declimatization　气候不适应
decomposed outcrop　风化露头
decorrelation　去相关,解相关
deep freeze　深度冻结
deep radar　深部探测雷达
deep sounding　深部探测
deep water　深层水
deep-penetrating radar　深部探测雷达
deep-sea core　深海岩芯
deficiency　缺陷
deflation　吹蚀
deflation basin　风蚀盆地
deflation hollow　风蚀洼地
deflation ripple　风蚀波痕
deformable aquifer　可形变含水层
deformation　变形
deformation observation　变形观测
deformation rate　变形速率
deformation till　变形冰碛
deformed ice　变形冰
deforming till　变形冰碛
degenerated form　退化类型
degenerated state　退化状态
degeneration　退化
deglacial　冰消期
deglaciation　冰消作用,冰消期
deglaciation landscape　冰川退缩景观
degradation　退化,剥蚀,降低

degradation polygon　退化多边形土
degraded ecosystem　退化生态系统
degraded forest　退化林
degraded land　退化土地
degree-day　度日
degree-day factor　度日因子
degree-day model (DDM)　度日模型
degree-hour　度时
dehydration by freezing　冻干,冻结脱水
deicer　防冰器,除冰器
deicing (=deice)　除冰
dekad　十天,旬
delayed drainage　延迟给水,滞后给水
delayed index　延迟指数
delayed strength　延滞强度
dell　小峡谷,小溪谷
delta moraine　三角洲冰碛
deluge　洪水(古)
dendritic crystal　枝状晶体,戟状晶体(雪花的一种)
dendritic glacier　树枝状冰川
dendritic snow crystal　枝状雪花晶体,枝状雪晶
dendroclimatology　树轮气候学
dendroecology　树轮生态学
dendrogeomorphology　树轮地貌学
dendroglaciology　树轮冰川学
dendrograph　树轮测定仪
denitrifying bacteria　脱氮细菌
densitometer　密度计
density current　密度流

density gradient 密度梯度
denudation 剥蚀,剥蚀作用
denudation landform 剥蚀地形
denudation plane 侵蚀面
departure 偏差,距平
depegram 露点图
depolarization 去极化
deposited moraine 堆积冰碛
deposition nucleus 凝华核
deposition velocity 沉积速度
depositional process 沉积过程
depositional rate 沉积速率
depositional sequence 沉积层序
depressed area 沉陷区,沉降区
depression 低〔气〕压,洼地,低地
depression belt 低压带
depression detention 填洼
depression glacier 洼地冰川
depth contour 等深线
depth datum 深度基准面
depth hoar 深霜
depth marker 深度标记
depth signal pole 水深信号杆
derived community 衍生群落
Derung Nationality 独龙族
desalinization 脱盐
desert perennial 荒漠多年生植物
desert soil 荒漠土壤
desert zone 荒漠带
desertification reversion 沙漠化逆转
desertuctive effect 破坏作用
desertuctive sampling 破坏性取样
desertuctive species 破坏种

desiccation crack 干裂缝
desiccation polygon 干裂多边形
design freezing index 设计冻结指数
design thawing index 设计融化指数
desquamation 剥蚀作用
desulfurication 脱硫化作用
detachment failure 剥离性滑塌
detection limit 检测限
deteriorated area 荒芜区
deterioration 恶化
deterministic hydrologic model 确定性水文模型
detrital 由岩屑形成的
dew 露
dew point 露点
diamicton 混杂堆积
diamond dust 钻石尘(晴天降雪)
diatom ooze 硅藻软泥
diatomite 硅藻土
diazotroph 固氮生物
Dicke radiometer 狄克式辐射计
dielectric constant 介电常数
dielectric profiling (DEP) 介电特征剖面
differential frost heave 不均匀冻胀,差异性冻胀
differential thaw settlement 差异性融沉
differentiation positioning 差分定位
diffluence glacier 分流冰川

diffraction 衍射
diffuse ice edge 扩散冰外缘
diffuse reflection 漫反射
diffuse sky radiation 漫射天空辐射
diffusion 扩散,漫射
diffusion coefficient 扩散系数,漫射系数
diffusion flux 扩散通量
diffusion transfer 扩散转移
diffusive resistance 扩散阻力
digital earth 数字地球
Digital Elevation Model (DEM) 数字高程模型
digital graphic processing 数字图形处理
digital image 数字图像
digital image processing 数字图像处理
digital map 数字地图
digital mapping 数字制图
digital mosaic 数字镶嵌
digital ortho image 数字正射图像
digital orthophoto map (DOM) 数字正射影像图
digital photogrammetry 数字摄影测量
digital surface model (DSM) 数字表面模型
digital-to-analog (D/A) conversion 数—模转换
digitization 数字化
digitized image 数字化图像
dihedral reflector 双面角反射器

dilation crack 张裂缝
dilation crack ice 张裂缝冰
diluvial period 洪积期
diluvial soil 洪积土
diluvium 洪积物
dimethyl sulphide (＝sulfide) (DMS) 二甲基硫
dimensionless parameter 无量纲参数
dimensionless unit hydrograph 无量纲单位线(水文)
dimethyl sulfide (DMS) 二甲基硫
dimethyl sulfoxide (DMSO) 二甲基硫亚砜
dimethylamine 二甲基氨
direct runoff 地表径流,直接径流
dirt cone 污化冰锥
dirty band 污化层,污化带
dirty ice 污化冰
discharge 流量
discharge area 排泄区
discharge capacity 排水能力
discharge flume 测流槽
discharge hydrograph 流量过程线
discharge recession 退水过程
discharge table 水位流量关系
discoid 盘状(冰碛石)
discordant intrusion 不整合侵入
Discovery 发现者卫星计划(美国)
discrete ordinate radiative transfer model (DISORT) 离散坐标辐射传输模型
discrimination 分馏

disequilibrium permafrost 非稳定态多年冻土
dispersion 色散,频散
dispersive capacity 扩散能力
displacement observation 位移观测
displacement of image 像点位移
dissolved inorganic carbon (DIC) 溶解态无机碳
dissolved organic carbon (DOC) 溶解态有机碳
dissolved phosphate 溶解磷酸盐
distance theodolite 测距经纬仪
distortion 畸变
distortion of projection 投影变形
diurnal cycle 日循环
diurnal range 日较差
diurnal variation 日变化
divide 分水(冰)岭
dog sled 狗拉雪橇
dome 雪帽,雪冠,冰穹
dome shaped dune 穹状沙丘
dome-shaped snow patches 穹状雪斑
dominant wind 盛行风
Donau Glaciation 多瑙冰期
Doppler frequency 多普勒频率
Doppler Orbitograph and Radio Positioning Intergrated by Satellite (DORIS) 多里斯系统
Doppler radar 多普勒雷达
Doppler sonar 多普勒声呐
double island arc 双岛弧
double-ripper 双雪橇

downcutting 下切作用
downfall 暴雨
downhill race 速降滑雪赛,滑降
downscaling 降尺度
downslope current sediment 坡水沉积物
downstream 下游
downwearing thermokarst 向下热侵蚀
drag coefficient 拖曳系数,阻力系数
drainage area 流域面积
drainage basin 流域(盆地)
drainage basin morphology 流域形态学
drainage density 河网密度
drainage divide 分水岭,流域分界线
drainage flow 泄流
drainage glacier 储水冰川
drainage map 水系图
dressed rock 羊背石
drift 冰碛,漂流
drift bed 坡积,漂碛
drift block 漂流石块
drift boulder 漂砾
drift chronology 冰碛年代学
drift clay 冰川泥砾
drift deposit 冰川沉积
drift detritus 漂移石屑
drift ice 流冰,漂流冰
drift ice foot 积雪斜坡
drift landform 冰碛地形

drift mound 冰碛丘
drift sand 流沙
drift scratch 冰碛擦痕
drift snow glacier 多年粒雪堆
drift stratigraphy 冰碛地层学
drift terrace 冰碛阶地
drift-barrier lake (=drift-dammed lake) 冰碛阻塞湖
drift-dammed lake (=drift-barrier lake) 冰碛阻塞湖
drifting snow 风吹雪，低吹雪（2 m 高度以下）
drifting station 浮冰漂流考察站（北冰洋）
driftless area 无冰碛区域
driven pile 打入桩
driven snow 吹雪
driving data 驱动数据
driving factor 驱动因子
driving force 驱动力
driving stress 驱动应力
drizzling fog 毛雨雾
drought 干旱
drought frequency 干旱频数
drought stress 干旱胁迫
drought-striken region 受旱地区
drowned valley 溺谷，沉没河谷
drumlin 鼓丘
drumlin field 鼓丘原，鼓丘群
drumlinization 鼓丘作用
drumlinoid drift tails 鼓丘状冰碛尾丘
dry bridge 旱桥

dry density 干密度
dry freeze 干冻
dry haze 干霾
dry ice 干冰
dry permafrost 干燥多年冻土
dry season runoff 枯季径流
dry snow 干雪
dry snow densification 雪密实化
dry snow zone 干雪带
Dry Valley 干谷（南极地名）
Dryas 仙女木期
dry-snow line 干雪线
dual medium 双重介质
dual-frequency sounder 双频测深仪
dump moraine 卸碛
dune 沙丘，雪丘
dune crest 沙脊
dune field 沙丘地
dune morphology 沙丘形态
dune ridge 沙垅
dune stabilization 沙丘固定
dust avalanche 干雪崩
dust layer 污化层
dust well 冰面融坑
dust haze 尘霾
dye tracer experiment 染色示踪试验
dynamic climatology 动力气候学
dynamic forecasting 动力预报
dynamic geodesy 动力大地测量学
dynamic glaciology 动力冰川学
dynamic loading effects 动载荷作用
dynamic model 动力学模型

dynamic Poisson's ratio 动泊松比
dynamic recrystalization 动力再结晶
dynamic viscosity coefficient 动力学黏性系数
dynamics 动力学
dystric geli-sandic primosols 酸性寒冻砂质新成土
dystric matti-gelic cambosols 酸性草毡寒冻雏形土

E

early frost 早霜
early frost hidden 黑霜
early snow 初雪,早雪
early-successional species 早期演替种
earth core 地核
earth crust 地壳
earth gravity model 地球重力场模型
earth hummock 土丘,冰土丘
earth light 地光,地球反照
earth mantle 地幔
Earth Observation System (EOS) 地球观测系统
earth orbit 地球轨道
earth radiation 地球辐射
Earth Radiation Budget Satellite (ERBS) 地球辐射收支测量卫星
Earth Resource Satellite (ERS) 中国地球资源卫星
Earth Resource Technology Satellite (ERTS) 地球资源技术卫星
earth tide 固体潮
earth-atmosphere radiation budget 地气辐射平衡
earth-atmosphere system 地气系统
earth-synchronous orbit 地球同步轨道
East Antarctic Ice Sheet (EAIS) 东南极冰盖
East Asian monsoon 东亚季风
East Greenland Current 东格陵兰海流
East Kamchatka Current 东勘察加海流
easy ice 易航冰区
eccentricity 偏心率
echo analysis 回波分析
echo delay time 回波延迟时间
echo sounder 回声测深仪,回声测深器
echo sounding 回声测深
echo-free zone (EFZ) 无回波带,

回波空白带
ecodistrict 生态区
ecological balance 生态平衡
ecological benefit 生态效益
ecological carrying capacity 生态承载力
ecological crisis 生态危机
ecological effect 生态效应
ecological equilibrium 生态平衡
ecological factor 生态因子
ecological function 生态功能
ecological geographic distribution 生态地理分布
ecological heterogeneity 生态异质性
ecological investigation 生态调查
ecological process 生态过程
ecological reserve 生态保护区
ecological restoration 生态恢复
ecological shelter 生态屏障
ecological structure 生态结构
ecological succession 生态演替
ecological survey 生态调查
ecological threshold 生态阈值
ecological water demand 生态需水
economic system 经济系统
ecosystem 生态系统
ecosystem diversity 生态系统多样性
ecosystem fragility 生态系统脆弱性
ecosystem function 生态系统功能
ecosystem integrity 生态系统完整性
ecosystem management 生态系统管理
ecosystem material cycle 生态系统物质循环
ecosystem model 生态系统模型
ecosystem process 生态系统过程
ecosystem service 生态系统服务
ecosystem stability 生态系统稳定性
ecosystem stress 生态系统胁迫
ecosystem structure 生态系统结构
ecotone 生态过渡带,生态交错带
ecotype 生态型
edaphic metabolism 土壤代谢
edaphic resources 土壤资源
edaphogenic succession 土壤发生演替
edaphon 土壤微生物(群体)
eddy conductivity 涡动传导率
eddy covariance (EC) 涡度相关
eddy diffusion coefficient 涡动扩散系数
eddy viscosity coefficient 涡黏性系数
edge detection 边缘检测
edge enhancement 边缘增强
edoma 冰楔复合体
Eemian Interglacial stage 埃姆间冰段
effective accumulated temperature 有效积温
effective evapotranspiration 有效

蒸散
effective flux 有效通量
effective freezing depth 有效冻结深度
effective head loss 有效水头损失
effective ice period (河、湖、海冰)有效结冰期
effective ice strength 有效冰强度
effective ice temperature 有效冰温
effective nocturnal radiation 有效夜间辐射
effective outgoing radiation 有效向外辐射
effective permeability 有效渗透率
effective porosity 有效孔隙度
effective precipitation 有效降水
effective radiation 有效辐射
effective radiation layer 有效辐射层
effective radiation temperature 有效辐射温度
effective roughness height 有效粗糙高度
effective roughness length 有效粗糙长度
effective sample size 有效样本数
effective section 有效断面, 有效截面
effective snowmelt 有效融雪
effective thermal conductivity 有效导热率
effective water-holding capacity 有效持水量, 田间持水量

egress 出口, 流出
eigenperiod 固有周期
eiscir 蛇形丘(爱尔兰)
ejectable radiosonde 弹射探空仪
El Niño 厄尔尼诺
El Niño/Southern Oscillation (ENSO) 厄尔尼诺/南方涛动
elastic coefficient 弹性系数
elastic modulus 弹性模量
electric conductivity raingauge 电导雨量计
electric cup anemometer 电传式风杯风速表
electric thermometer 电测温度仪
electrical conductivity (EC) 电导率
electrical hygrometer 电测湿度仪
electrical resistivity tomography (ERT) 电阻率成像[技术]
electrical substitution radiometer 电测辐射表
electrically scanned microwave radiometer (ESMR) 电子扫描微波辐射计
electrically-heated thermometer 电热式温度表
electric-capacity moisture meter 电容式含水量测定仪
electrochemical detector (ECD) 电化学检测器
electrochemical sonde 电化学探空仪(测臭氧用)
electrolytic hygrometer 电解式湿

度表
electrolytic thermometer 电解温度计
electromagnetic profiling 电磁法剖面勘探
electron spin resonance（ESR） 电子自旋共振
electronic psychrometer 电解干湿表
electronic tacheometer 电子速测仪
electronic temperature recorder 电子温度记录仪
electronic theodolite 电子经纬仪
electronic thermometer 电子干湿表
electropsychrometer 电测湿度计
electrostatic aerosol sampler 静电式气溶胶取样器
elevated temperature psychrometer 增温式干湿表
elevation head 位势水头
elevation point 高程点
ellipsoidal geodesy 椭球面大地测量学
elution processes 淋溶过程
eluvial facies 残积相
eluvial horizon 淋溶层
eluviation 淋溶作用
embacle （河流)碎冰群
embedded 镶嵌的,嵌入的
embryo esker 雏形蛇形丘
emergence flow （冰川)显出流
emergy analysis 能值分析
emergy transformity 能值转换率
emission 发射,排放

emission control standard 排放控制标准
emission reduction unit（ERU） 排放减量单位
emission scenarios 排放情景
emission source 发射源,排放源
emissivity 发射率,排放率
empirical orthogonal function （EOF） 经验正交函数
empirical unit hydrograph 经验单位线
empower 能值功率
enclosed basin 闭合流域
end moraine 终碛
endangered ecosystem 濒危生态系统
endogenetic succession 内因性演替
energy absorption 能量吸收
energy balance 能量平衡
energy balance model 能量平衡模型
energy budget 能量收支,能量平衡
energy exchange 能量交换
energy flux 能量通量
engineering geocryology 工程冻土学
engineering hydrology 工程水文学
engineering technogenic talik 工程融区
englacial 冰内的
englacial debris 冰内岩屑
englacial drainage 冰内水系
englacial drift 冰川内碛(旧称）
englacial material 冰内物质
englacial melt 冰内融化
englacial stream 冰内河流

englacial till 冰内碛
englacial water-table 冰内水位
enhanced greenhouse effect 增强温室效应
enhanced thematic mapper (ETM+) 增强型专题制图仪
environment impact assessment (EIA) 环境影响评价
environment impact statement (EIS) 环境影响评价
environmental abnormality 环境异常性
environmental evolution 环境演化
environmental factor 环境因子
environmental geochemistry 环境地球化学
environmental hazard 环境灾害
Environmental Protection Act (美国)环境保护法
Environmental Research Satellite (ERS) (美国)环境研究卫星
Environmental Satellite (ENVISAT) (极轨地球观测计划)环境卫星(欧空局)
environmental sensitive area (ESA) 环境敏感区
Eocene 始新世,始新世的
Eocene to Oligocene transition 始新世—渐新世气候转型
Eogene period 早第三纪
Eogene System 下第三系
eolian action 风成作用
ephmeral frozen ground 瞬时冻土
epigeneisis 后生作用
epigenetic ground freezing 后生冻结
epigenetic ice 后生冰
epigenetic ice wedge 后生冰楔
epigenetic permafrost 后生多年冻土
epigleyic geli-sandic primosols 表潜寒冻砂质新成土
equally tilted photography 等倾摄影
equatorial counter-current 赤道逆流
equi energy spectrum 等能量光谱
equiaccuracy chart 等精度〔曲线〕图
equidensen 等密度面
equidistant projection 等距投影
equilibrium line 平衡线
equilibrium line altitude (ELA) 平衡线高度
equilibrium permafrost 平衡态多年冻土
equilibrium rebound 均衡反弹
equivalent 等值,当量
equivalent hydraulic transmissibility 等效渗透系数
eroding force 侵蚀力
eroding stress 侵蚀应力
erosion 侵蚀,侵蚀作用
erosion basis 侵蚀基准面
erosion lake 侵蚀湖
erosion modulus 侵蚀模数
erosion rate 侵蚀速率
erosion ratio 侵蚀比率

erosion ridge 侵蚀脊
erosion surface 侵蚀面
erosion terrace 侵蚀台地
erosional gully 侵蚀沟
erosional landform 侵蚀地形
erosional stage 侵蚀期
erosivity 侵蚀力
erratic boulder (block) 漂砾
erratic index 漂砾指数
erratic raft 巨型漂砾
erratic till 漂砾碛
eruption type 火山喷发类型
escar(=eschar) 蛇形丘
escarpment 陡崖蛇形丘,马头丘
esker 蛇形丘(爱尔兰)
esker delta 蛇形丘三角洲
esker fan 蛇形丘扇
esker-like ridge 蛇形丘岭
Eskimo 爱斯基摩人
Eskimo dog 爱斯基摩狗
Esquimau 爱斯基摩语
estuarine delta 三角湾状三角洲
estuarine lake 河口湖
estuary 三角湾
estuary coast 三角湾海岸
etang 滩积内陆湖
etch 刻蚀,侵蚀
etch pit 刻蚀坑
etched figure 刻蚀形态
etched groove 刻蚀沟
etched pothole 刻蚀锅穴
etching 刻蚀作用
etching figure 刻蚀形态

etchplain （冰川)侵蚀平原
ethnoecology 民族生态学
Etna volcano 埃特纳火山
etrier 绳梯
eukaryotes 真核生物
eulittoral zone 潮间带
eupatheoscope 热耗仪
Euro-Asian Continental Grassland Transect (EACGT) 欧亚大陆草地样带
European Remote-Sensing Satellite (ERS) 欧洲遥感卫星
European Space Agency (ESA) 欧洲空间局
eurythermal 广温性的
eurytopic species 广适种
eutrophication 富营养化
evaporation 蒸发
evaporation flux 蒸发通量
evaporation frost 蒸发霜
evaporation rate 蒸发速率
evaporation tank 大型蒸发皿
evaporimeter 蒸发仪
evaporograph 蒸发计
evapotranspiration 蒸散发,蒸散
evergreen broadleaf forest (EBF) 常绿阔叶林
evergreen needleleaf forest (ENF) 常绿针叶林
evolutionary adaptability 进化适应性
Ewenki Nationality 鄂温克族
excavation 掘蚀,淘蚀
excess ice 过量冰,超饱和冰

excess moisture 过剩水分,超渗水
exhaustible resources 可耗竭资源
exhaustive search 穷举搜索法
existing glacier 现代冰川
exogenetic succession 外因性演替
expanded-foot glacier 宽尾冰川
expansibility 膨胀率
expansion 膨胀
expansion cooling 膨胀冷却
exploitation management 掠夺式经营
exposure 暴露,暴露度
exposure age 暴露年龄
exposure assessment 暴露评估
exposure unit 暴露单元
ex-situ conservation 迁地保护
extending flow 伸张流
extensification 粗放化
extensive management 粗放经营
external environment 外界环境
external factor 外界因子
external forcing function 外界驱动作用
external water circulation 外水分循环
extinct volcano 死火山
extinction 消光
extraglacial deposit 冰川外围沉积
extraglacial drainage 冰川外围水系
extra-tropical cyclone 温带气旋
extreme climate 极端气候
extreme climate event 极端气候事件
extreme weather event 极端天气事件
extreme-frost susceptibility 极高冻胀敏感性
extrusion flow 挤出流
extrusive ice 喷出冰

F

facetted spur 三角嘴,削切坡,截切山嘴
facies (沉积)相
facies analysis (沉积)相分析
facies sequence 相序
Fahrenheit thermometer 华氏温度表
fall frost 秋霜冻
fallstreak 雨〔雪〕幡
false color image 假彩色图像
fan glacier 扇状冰川
fan tongue 扇状舌
fan-shaped rock glacier 扇形石冰川
far infrared 远红外
farming-pastoral zone 农牧交错带

farmland-protective forest 农田防护林
fast Fourier transform (FFT) 快速傅里叶变换
fast ice 固定冰(海冰)
fast ice boundary 固定冰边缘(海冰)
fast-response ozone detector 快〔速〕响应臭氧检测器
fathogram 水深图
fault coast 断层海岸
favourable temperature 适宜温度
feathery crystal 羽状晶体
feature extraction 特征提取
feature point 特征点
feature tracking 特征识别,特征追踪
feedback 反馈
feedback loop 反馈环
fen 草本沼泽
fen peat 低沼泥炭
fenced plot 围栅样地
feng Yun (FY) meteorological satellite "风云"气象卫星(中国)
fenland 沼泽地,干沼泽
fenny 沼泽的,生于沼泽地带的
fern 蕨,蕨类植物
ferralarch 自然演替系列
ferredoxin 铁氧化还原蛋白
Ferrel cell 费雷尔环流圈
fertility rate 人口出生率,生育率
fetch 风区,风区长度,吹程
fibric histi-permagelic gleyosols 纤维有机多年冻结潜育土
fibric organic cryosol 纤维有机寒冻土
fibric soil materials 纤维有机土壤物质
fibristels 纤维有机寒冻土
fibrous ice 纤维状冰
field mapping 野外填图
figure skate 花样滑冰鞋
figure skating 花样滑冰
Fildes Peninsula 菲尔德斯半岛(南极)
filling barometer 充液气压计
film crust 薄冰壳
film moisture (= film water) 薄膜水
film water migration 薄膜水迁移
fine grain snow 细雪
fine particles 细颗粒
fine-grained soil 细粒土
finger rafted ice 指状重叠冰
finite difference method 有限差分法
finite element method 有限元法
firn 粒雪
firn bank 粒雪堤
firn basin 粒雪盆
firn field 粒雪原,粒雪场
firn ice 粒雪冰
firn limit 粒雪界限
firn line 粒雪线
firn zone 粒雪区
firncrash 粒雪破碎(旧称)

firnification 粒雪化〔过程〕
firnpush 粒雪崩落(旧称)
firnspiegel 粒雪镜冰(旧称)
firnstoss 粒雪崩落(旧称)
First Frost 霜降(节气)
first snowfall 初雪
first year ice (海冰的)一年冰
first-ice date 初冰日
fission track dating (FTD) 裂变径迹测年
fission track geochronology 裂变径迹地质年代学
fissure volcano 裂隙式火山
fixed cistern barometer 定槽式气压表
fixed dune 固定沙丘
fixed platform 固定平台
fixed ship station 固定船舶站
fjall 冰蚀高原
fjeld 冰蚀高原
fjeldbotn 冰蚀冰斗
fjell 冰蚀沼地
fjord strait 峡湾式海峡
fjord stream 峡湾河
fjord topography 峡湾地形
fjord valley 峡湾谷
fjord (= fiord = fjard = forde) 峡湾
fjord〔type〕coast 峡湾式海岸
flag cloud 旗云
flake 雪片
flank moraine 侧碛
flash flood 山洪,突发性洪水
flash runoff 暴涨径流
flask sampling 瓶采样
flat-floored valley 平底谷
flaw lead (海冰)裂缝水道
flaw polynya 裂缝冰间湖
float type raingauge 浮筒雨量器
floating 漂浮的,浮游的
floating beacon 浮标
floating ice 浮冰
floating pan 漂浮蒸发皿
floating plant 浮水植物
floating radiosonde 漂浮探空仪
floe 浮冰块
floe belt 浮冰带
floe ice 浮冰
floeberg 浮冰丘,小冰山
floebit 浮冰块
flood alleviation 防汛
flood basin deposit 河漫盆地沉积
flood peak 洪峰
flood plain deposit 河漫滩沉积
flood warning 洪水预警
flood-initiation threshold 溃决临界条件
flora 植物区系
flow law 流动定律
flow line 流线
flow modulus (= runoff modulus) 径流模数
flow net 流网
flow splitting 分流
flow till 流碛
flow velocity 流速

flowmeter 流速仪
flurry 小阵雪
fluvaquentic aquorthels 冲积含水正常寒冻土
fluvaquentic fibristels 冲积含水纤维有机寒冻土
fluvaquentic haplorthels 冲积含水弱育正常寒冻土
fluvaquentic hemistels 冲积含水半腐有机寒冻土
fluvaquentic historthels 冲积含水有机正常寒冻土
fluvaquentic sapristels 冲积含水高腐有机寒冻土
fluventic haplorthels 冲积新成弱育正常寒冻土
fluventic historthels 冲积新成有机正常寒冻土
fluvial action 流水作用
fluvial sediment 河流沉积
fluvioglacial（＝aqueoglacial＝glacioaqueous＝glaciofluvial）冰水的
fluvioglacial accumulation 冰水堆积
fluvioglacial delta 冰水三角洲
fluvioglacial deposit 冰水沉积
fluvioglacial drift 冰水冲积
fluvioglacial erosion 冰水侵蚀
fluvioglacial lake 冰水湖
fluvioglacial terrace 冰水阶地
fluviothermal erosion 流水热力侵蚀
flux gradient method 通量梯度观测法
fluxion structure 流纹构造
foam crust 波状雪面
foehn（＝föhn） 焚风
fog 雾
fog crystal 雾晶
fog deposit 雾沉积
fog visiometer 雾天能见度仪
fog water collector 雾滴收集器
foggara 坎儿井
fog-gauge（＝fogmeter） 雾量计
foliated ice 片状冰，叶状冰
folic soil materials 落叶有机土壤物质
folistels 落叶有机寒冻土
folistels haplorthels 落叶弱育正常寒冻土
folk culture 民俗文化
folk custom 民俗
follistic molliturbels 落叶暗沃扰动寒冻土
follistic mollorthels 落叶暗沃正常寒冻土
follistic umbriturbels 落叶暗瘠扰动寒冻土
follistic ummrorthels 落叶暗瘠正常寒冻土
food chain 食物链
footprint 足迹
forage 采食牧草
Forbes band （冰面上的）福布斯带
forcing factor 强迫因子
fore trough 前槽

forest biomass 森林生物量
forest coverage 森林覆盖率
forest ecosystem 森林生态系统
forest fire 林火
forest hydrology 森林水文学
forest meteorology 森林气象学
forest microclimate 森林小气候
forest steppe 森林草原
forest swamp 森林沼泽
form factor （流域）形状系数
formate 甲酸盐
forward scattering spectrometer 前向散射光谱仪
fossil dune 古沙丘
fossil form 古形态
fossil fuel 化石燃料
fossil ice 古冰
fossil periglacial phenomenon 古冰缘现象
fossil permafrost 古冻土
fossil pingo 古冰丘
fossil rock glacier 古石冰川
fossil soil 古土壤
four-man bobsled 四人座有舵雪橇
fractional vegetation cover 植被盖度
fractionation 同位素分馏
fracture toughness 断裂韧性
fracture zone 破裂带
fracturing 破裂
fragmentation 破碎〔作用〕
fragmented landscape 破碎景观
fragmic cryogenic fabric 冷生片状聚合微结构
fragmoidal cryogenic fabric 冷生似片状聚合微结构
frash ice 碎冰
Fraunhofer diffraction 夫琅和费衍射
frazil crystal 冰针冰晶
frazil ice （水内的）冰针冰
frazilization 冰针现象
free aquifer 潜水层
free atmosphere 自由大气
free surface wave 自由表面波
free water 〔土壤〕自由水
Free-Air Carbon Dioxide Enrichment (FACE) 自由大气二氧化碳富集
freestyle skiing 自由式滑雪
freestyle snowboarding 自由式单板滑雪
freeze 冻结
freeze dewatering 冻结脱水
freeze drying 冷冻干燥
freeze spray 冻结状飞沫
freezeback around wells 井壁回冻
freezeback foundation soils 地基土回冻
freezeback period 回冻期
freezeback piles 桩基回冻
freeze-free period 非冻结期
freeze-level chart 冻深图
freeze-thaw action 冻融作用
freeze-thaw boundary 冻融边界
freeze-thaw cycle 冻融循环

freeze-thaw debris 冻融岩屑
freeze-thaw extrusion structures 冻融挤压结构
freeze-thaw interface 冻融界面
freeze-thaw pattern 冻融型式
freeze-thaw phenomena 冻融现象
freeze-thaw processes 冻融过程
freeze-thaw rock falls 冻融岩崩
freeze-thaw weathering 冻融风化
freeze-up 封冻,封河
freeze-up date 封冻日期
freeze-up period 封冻期
freezing borehole 冻结钻孔
freezing carrier 冷冻船
freezing curve 冻结曲线
freezing damage 冻害
freezing degree-day 冻结度日,负积温
freezing drizzle 毛毛冻雨
freezing fog 冻雾,雾凇
freezing fringe 冻结缘
freezing front 冻结界面
freezing hardiness 冻结硬度
freezing hole 冻结孔
freezing index 冻结指数
freezing injury 冻伤
freezing interface 冻结界面
freezing level 冻结高度
freezing mechanism 冻结机理
freezing nuclei spectrum 冻结核谱
freezing nucleus 冻结核
freezing period 结冰期,冻结期
freezing phase 冻结相
freezing point 冰点,冻结温度
freezing point depression 冻结点降低,冰点降低
freezing precipitation (=freezing rain) 冻雨
freezing pressure 冻结压力
freezing process 冻结过程
freezing rain (=freezing precipitation) 冻雨
freezing season 冻季
freezing soil 正冻土
freezing spray 冻结飞沫
freezing temperature 冻结温度
freezing zone 冻结带,冻结区
freezing-induced pressure 冻结诱发压力
freezing-level chart 结冰高度图
freezing-point depression 冰点降低
frequency bandwidth 频率带宽
frequency drift 频率漂移
frequency error 频率误差
frequency filtering 频率域滤波
frequency response 频率响应函数
frequency response model 频率响应模式
frequency spectrum 频率谱
frequency-modulated continuous-wave (FMCW) 调频连续波
fresh snow cover 新积雪
fresh water lake 淡水湖
friable permafrost 松散多年冻土
frigid blasts 凛冽寒风
frigid weather 严寒天气

frigid zone　严寒地区
frigofuge　畏寒植物,避寒植物
frigorideserta　冻原荒漠群落
frigorigraph　冷却计
frigorimeter　冷却仪
fringe glacier　边缘冰川
frith　河口
front　锋
front of snowmelt　融雪锋面
frontal aprons　冰前沉积
frontal circulation　锋面环流
frontal cyclone　锋面气旋
frontal lifting　锋面抬升
frontal moraine　终碛
frontal structure　锋面结构
frost　霜,结冰
frost action　冻融作用
frost belt　冻结带
frost blister　穹状冻胀丘
frost boiling　冻融翻浆,土冻沸作用
frost bound　冻结的,结霜的,冰冻的
frost buildup　积霜
frost bulb　冻结盘
frost churning　冻融泥流作用,冻融扰动作用
frost cleft　冻裂
frost climate　寒冷气候
frost cracking　冻结裂缝,寒冻开裂
frost creep　冻融蠕动
frost dam　冻堤
frost damage　霜冻,霜害

frost day　霜日
frost depth　冻结深度
frost detector　霜检测器
frost disaster　霜灾
frost feather　霜羽
frost fissure　冻裂缝
frost fissure polygons　冻裂多边形
frost fog　霜雾
frost-free day　无霜日
frost fringe　冻结缘
frost front　冻结锋
frost hazard　霜害,冻害
frost haze　冻霾
frost heave　冻胀
frost heave phenomena　冻胀现象
frost heave rate　冻胀速率
frost heave stress　冻胀应力
frost heaving　冻胀作用
frost heaving force　冻胀力
frost heaving ratio　冻胀率
frost impermeable belt　冻结防渗带,冻结防渗芯墙
frost injury　霜害,冻害
frost jacking　冻拔作用
frost line　霜线,冰冻线
frost mound　冻胀丘
frost number　冻结数
frost penetration　冻结面下移
frost period　冻结期
frost plants (= cryoplankton)　冰藻
frost point　霜点
frost prevention　霜冻预防

frost protection 霜冻防护
frost riving 寒冻风化,冻裂
frost scaling 霜冻剥落
frost scar 冻疤,冰冻痕
frost season 霜期,冻结期
frost shattering 冻融崩解作用,冻碎作用,冻劈作用
frost shifting 冻融滑移
frost smoke 霜雾
frost snow 冰晶簇
frost soil 寒冻土
frost sorting 冻融分选
frost splitting 冻裂,冰裂作用,融冻崩解作用
frost stirring 冻扰作用
frost susceptibility 冻胀敏感性
frost table 冻结面
frost thrusting 冻裂突起
frost tilting 冻倾作用
frost tolerance 耐寒性
frost tube 管式冻土深度探测器,管式冻土器(达尼林冻土器)
frost weather 霜冻天气
frost weathering 寒冻风化
frost wedging 冻劈作用,寒冻楔入过程
frost zone 冻结区
frost-active soil 冰冻作用土层
frostbite 冻伤,霜害
frost-crack polygon 冻裂多边形,冻龟裂
frost-cracking 冻裂作用
frost-free period 无霜期

frost-free season 无霜季节
frost-heave mound 冻胀丘
frosting 结霜
frostless season 无霜期
frostless zone 无霜带,无霜区
frost-lifting 冻拔
frost-point hygrometer 霜点湿度计
frost-point technique 霜点法
frost-point thermometer 霜点温度表
frost-shattered debris 冻裂碎岩屑
frost-susceptible classification 冻胀敏感性分类
frost-susceptible ground 冻胀敏感性土
frost-susceptible soil 冻胀敏感性土,冻结敏感性土
frost-thaw feature 冻融特征
frost-weathering action 寒冻风化作用
frost-weathering process 寒冻风化过程
frosty 霜冻的,结霜的
frozen carbon 冻结碳
frozen debris 冻结岩屑
frozen depth 冻结深度
frozen earth 冻土
frozen fog 冻雾
frozen ground 冻土
frozen ground blasting 冻土爆破
frozen ground ecology 冻土生态学
frozen ground engineering geology 冻土工程地质学
frozen ground hydrogeology 冻土

水文地质学
frozen ground mechanics 冻土力学
frozen ground rheology 冻土流变学
frozen precipitation 固态降水
frozen sediment 冻结沉积物
frozen snow crust 冻结雪壳
frozen soil 冻土
frozen soil apparatus 冻土深度探测器,冻土器
frozen soil classification 冻土分类
frozen ground phenomena 冻土现象
Fuji volcano 富士火山
fully-focused Synthetic Aperture Radar (fully-focused SAR) 全聚焦合成孔径雷达
fusion 熔化,融合

G

gage height 水尺水位
gaging station 水位站
gale 8级以上大风
Galvanic ozone monitor 伽伐尼臭氧监测仪
Gamburtsev Subglacial Mountains (GSM) 甘伯采夫冰下山脉
Garrulax henrici (Henry's Laughingthrush) 灰腹噪鹛
gas chromatograph 气相色谱仪
gas chromatography-mass spectrometer (GC-MS) 气相色谱－质谱仪
gas density balance 气体密度天平
gas state equation 气体状态方程
gaseous diffusional separation 气体扩散分离〔法〕
gaseous phase 气相
gaseous pollutant 气体污染物
gaseous pollution 气体污染
gauge catch ratio 雨量筒捕捉率
Gazetteer Period 方志时期
geocoded terrain corrected (GTC) 地形校正编码
gelic aqui-orthic halosols 寒冻潮湿正常盐成土
gelic cambosols 寒冻雏形土
gelic hapli-orthic spodosols 寒冻简约正常灰土
gelifluction 融冻泥流作用
gelifluction bench 融冻泥流条形台地
gelifluction deposit 融冻泥流沉积
gelifluction lobe 融冻泥流舌
gelifluction sheet 融冻泥流被,融冻泥流席

gelifluction slope 融冻泥流坡
gelifluction step 融冻泥流台阶，融冻泥流坡坎
gelifluction terrace 融冻泥流阶地
gelifraction 冰冻崩解，冻劈作用
gelisol 寒冻土
gelisol mapping 寒冻土制图
gelisol order 寒冻土纲
gelivation 寒冻作用
General Circulation Model（GCM） 大气环流模式
gentle breeze 3级以下微风
gentoo penguin（*pygoscelis papua*） 金图企鹅，巴布亚企鹅，绅士企鹅
geobiont 土壤生物
geobotanical regionalization 植被区划
geobotany 地植物学
geochemical cycle 地球化学循环
geochemical environment 地球化学环境
geochemist 地球化学家
geochemistry 地球化学
geochronology 地质年代学
geochronometric scale 地质年表，地质编年表
geochronometry 地质年代测定〔法〕
geocryology 冻土学
Geodetic and Earth Orbiting Satellite（GEOS） 地球轨道测地卫星
geodetic database 大地测量数据库
geodetic datum 大地基准
geodetic measurement 大地测量法
Geodynamics Experimental Ocean Satellite（GEOS） 地球动力学实验海洋卫星
geoecology 地生态学
geoelectrical resistivity measurement 地电阻率法
geoelectrical sounding 地电探测〔法〕
geographer 地理学家
geographic information systerm（GIS） 地理信息系统
geographic meridian 地理子午线
geographical environment 地理环境
geographical environment determinism 地理环境决定论
geographical landscape 地理景观
geographical name index 地名索引
geography 地理学
geoid 地球体，大地水准面
geologic hazard 地质灾害
geologic thermometer（= geothermometer） 地质温度计
geologic time 地质时期，地质年代
geologic time scale 地质年代尺度
geological age 地质时代
geological chemistry 地质化学
geological climate 地质气候
geological period 地质时期
geology 地质学
geomagnetic activity index 地磁活动指数

geomagnetic excursion 地磁漂移
geomagnetic meridian 地磁子午圈
geomagnetic pole 地磁极
geometric calibration 几何校准
geometric correction 几何纠正
geometric distortion 几何畸变
geomorphic accident 地貌事件
geomorphic cycle 地貌旋迴
geomorphic development 地貌发育
geomorphic history 地貌历史
geomorphic snow line 地形雪线
geomorphic unit 地貌单元
geomorphochronology 地貌年代学
geomorphogenesis 地貌成因
geomorphogeny 地貌成因学,地形发生学
geomorphologic(al) guide 地貌标志
geomorphologic(al) sequence 地貌序列
geomorphological map 地貌图
geomorphological process 地貌过程
geomorphologist 地貌学家
geomorphology 地貌学
geomorphy 地貌
geophysical survey 地球物理调查
geophysicist 地球物理学家
geophysics 地球物理学
geophytia 地面植物群落
geopotential height 位势高度
geopotential meter (gpm) 位势米
georeferencing 地理配准
geoscience 地球科学
geoscience laser altimeter system (GLAS) 地球科学激光测高系统
Geostationary Meteorological Satellite (GMS) 地球静止气象卫星
Geostationary Operational Environmental Satellite (GOES) 地球静止业务环境卫星(美国)
Geostationary Operational Meteorological Satellite (GOMS) 地球静止业务气象卫星
geostrophic balance 地转平衡
geostrophic force 地转偏向力
geo-synchronous satellite 地球同步卫星
geosystem 地理系统
geotechnological regionalization 地学工程技术区划
geothermal activity 地热活动
geothermal anomaly 地热异常
geothermal condition 地热条件
geothermal data 地热资料
geothermal gradient 地热梯度
geothermal heat 地热
geothermal heat flux 地热通量
geothermal talik 地热融区
geothermics 地热学
geothermometer (= geologic thermometer) 地质温度计
giant floe 巨型浮冰
giant-grained 巨粒的
giant-granite 花岗岩巨砾
gill 沟壑,涧流

glacial geological culture　冰川地质文化
glacial geology　冰川地质学
glacial geomorphic activity　冰川地貌作用
glacial geomorphology　冰川地貌学
glacial-interglacial change　冰期一间冰期变化
glacial-interglacial cycle　冰期一间冰期旋迴
glacial-interglacial transition　冰期一间冰期过渡
glacialism　冰川理论（旧称）
glacialization　冰川作用
glacially eroded depression　冰蚀洼地
glaciated area　冰川作用〔过的〕区
glaciated landform　冰川侵蚀地貌
glaciated plain　冰川侵蚀平原
glaciated rock　冰蚀岩
glaciated valley　冰蚀谷
glaciation　冰川作用
glaciation evidence　冰川作用证据
glaciation level　冰川作用程度
glaciation limit　冰川作用下限
glacic anhyorthels　厚冰层脱水正常寒冻土
glacic anhyturbels　厚冰层脱水扰动寒冻土
glacic aquiturbels　厚冰层含水扰动寒冻土
glacic aquorthels　厚冰层含水正常寒冻土
glacic argiorthels　厚冰层黏化正常寒冻土
glacic folistels　厚冰层落叶有机寒冻土
glacic haplorthels　厚冰层弱育正常寒冻土
glacic haploturbels　厚冰层弱育扰动寒冻土
glacic historthels　厚冰层有机正常寒冻土
glacic histoturbels　厚冰层有机扰动寒冻土
glacic molliturbels　厚冰层暗沃扰动寒冻土
glacic mollorthels　厚冰层暗沃正常寒冻土
glacic organic cryosol　厚冰层有机寒冻土
glacic psammorthels　厚冰层粗粒正常寒冻土
glacic psammoturbels　厚冰层粗粒正常扰动寒冻土
glacic umbriturbels　厚冰层暗瘠扰动寒冻土
glacic umbrorthels　厚冰层暗瘠正常寒冻土
glacier　冰川
glacier ablation　冰川消融
glacier abrasion　冰川磨蚀
glacier advance　冰川前进
glacier adventure　冰川探险
glacier age（＝ice age）　冰期
glacier airplane tour　冰川航空旅游
glacier alluvion　冰川冲积层

glacier amphitheater 粒雪盆
glacier anticyclone 冰川反气旋
glacier anticyclone theory 冰川反气旋学说
glacier area 冰川区
glacier avalanche 冰崩
glacier band 冰川带
glacier basin (＝glacial basin) 冰川流域
glacier basin 粒雪盆,冰川流域
glacier bed 冰床
glacier boulder 冰川漂砾
glacier boulder clay 冰川泥砾
glacier boundary 冰川边界
glacier breecia 冰川角砾
glacier breeze 冰川风
glacier cableway 冰川索道
glacier canal 冰川河道
glacier canyon 冰川峡谷
glacier carved valley (＝glacier valley) 冰蚀谷
glacier cascade (串珠状)冰瀑布
glacier cave 冰洞
glacier change 冰川变化
glacier chronology 冰川年代学
glacier chute 冰川刻槽
glacier circular tour 冰川环程旅行
glacier cirque 冰斗
glacier clay 冰川黏土
glacier cliff 冰崖
glacier conglomerate 冰砾岩
glacier crevasses 冰川裂隙
glacier cruise 冰川航海观光
glacier cultural landscape 冰川文化景观
glacier dammed lake 冰川堰塞湖
glacier debris flow 冰川泥石流
glacier delta 冰川三角洲
glacier denudation 冰蚀
glacier deposit 冰川沉积物
glacier deposition 冰川沉积作用
glacier diffluence 冰川分流
glacier diffluence pass 冰川分流垭口,冰川越过山口
glacier discharge 冰川冰流量
glacier dispersion 冰川消退,冰川退消
glacier divide 分冰岭
glacier drag 冰川拖曳
glacier drift 冰碛
glacier drift boulder 冰川漂砾
glacier drift cliff 冰碛悬崖
glacier dynamics 冰川动力学
glacier episode 冰川幕
glacier epoch 冰期
glacier epoch origin theory 冰期起源假说
glacier eroded depression 冰蚀洼地
glacier eroded trough 冰川刻蚀槽
glacier erosion 冰川侵蚀〔作用〕
glacier erosion cycle 冰蚀循环
glacier erosion lake 冰蚀湖
glacier erosion surface 冰川侵蚀面
glacier erratic 冰川漂砾
glacier eustasy 冰期海面升降运动

glacier eustatic change 冰川导致的海平面变化
glacier evolution 冰川演化
glacier excursion 冰川考察
glacier expedition 冰川探险
glacier explorer 冰川探险者
glacier express tour 冰川快车旅游
glacier facies 冰川相,冰川带
glacier fall（＝glacier cascade） 冰瀑布
glacier feature 冰川特征
glacier fissure 冰裂隙
glacier flood 冰川洪水
glacier flora 冰川植物区系
glacier flour 冰川岩粉
glacier fluctuation 冰川波动
glacier fluting 冰刻蚀凹槽
glacier fluvial landform 冰川河流地貌
glacier front 冰川前缘
glacier funnel 冰面漏斗
glacier geological museum 冰川地质博物馆
glacier geological park 冰川地质公园
glacier gorge 冰蚀峡谷
glacier gravel 冰川砾石
glacier grinding 冰川碾磨作用
glacier groove 冰刻沟,冰蚀沟
glacier guide 冰川导游
glacier hanging valley 冰成悬谷,冰蚀悬谷
glacier helicopter tour 冰川直升机旅游
glacier high （大冰盖区）冰川高压
glacier hike 徒步冰川旅行
glacier history culture 冰川历史文化
glacier holiday resort 冰川度假胜地
glacier horn 冰川角峰
glacier hotel 冰川旅馆
glacier hydraulics 冰川水力学
glacier ice 冰川冰
glacier interstade 冰期间冰段
glacier inventory 冰川编目
glacier isostasy 冰川地壳均衡说
glacier isostatic adjustment 冰川地壳均衡调整
glacier journey（＝glacier trip） 冰川旅行
glacier karst 冰川喀斯特
glacier kayaking tour 冰川皮划艇旅游
glacier lake 冰川湖
glacier lake deposit 冰川湖沉积
glacier lake inventory 冰湖编目
glacier lake outburst flood（GLOF） 冰湖溃决洪水
glacier landform 冰川地形
glacier landscape 冰川景观
glacier levee 冰砾堤
glacier lobe 冰舌
glacier mammillation 冰川磨蚀作用
glacier marginal lake 冰川边缘湖

glacier mass balance	冰川物质平衡
glacier (mass) budget	冰川(物质)收支
glacier maximum	冰盛期
glacier meal	冰川岩粉
glacier meltwater	冰川融水
glacier meltwater pool	冰面湖泊
glacier milk	冰川乳
glacier mill	冰川竖井,冰磨房
glacier moraine	冰碛垄
glacier morphology	冰川形态学
glacier moulin	冰川竖井
glacier movement	冰川运动
glacier mud	冰川泥
glacier museum	冰川博物馆
glacier origin	冰川成因
glacier oscillation	冰川脉动
glacier outwash	冰水沉积物
glacier overdeeping	冰川过量掘蚀作用
glacier park	冰川公园
glacier pavement	冰溜面
glacier period	冰期
glacier phase	冰阶〔段〕,冰川相
glacier plain	冰成平原
glacier planation	冰川夷平作用
glacier ploughing	冰川刨掘作用
glacier plucking	冰川掘蚀作用
glacier polish	冰川磨光面
glacier process	冰川过程
glacier protection	冰川保护
glacier quarrying	冰川拔蚀作用
glacier rebound	冰盖消退后地壳反弹
glacier recession	冰川后退
glacier record	冰川记录
glacier refuge	冰川保护区
glacier regime	冰川状态
glacier relic	冰川遗迹
glacier relic flora	冰期残存植物区系
glacier relicts	冰期子遗
glacier replenishment	冰川补给
glacier resource	冰川资源
glacier response time	冰川响应时间
glacier retreat	冰川退缩
glacier rock	冰成岩
glacier route	冰川游览线路
glacier runoff	冰川径流
glacier sapping	冰川挖掘作用
glacier scouring	冰擦作用
glacier scratch	冰川抓挖作用
glacier sculpture	冰川刻蚀
glacier sculpture terrain	冰川刻蚀地形
glacier sculptured landscape	冰川刻蚀景观
glacier sediment	冰川沉积物
glacier shrinkage	冰川萎缩,冰川退缩
glacier sightseeing	冰川观光
glacier silt	冰川粉砂土
glacier ski resort	冰川滑雪胜地
glacier ski tour	冰川滑雪旅游

glacier skiing 冰川滑雪
glacier slide 冰川滑动
glacier snout (＝glacier tongue)
　冰舌
glacier-snow ablation 冰雪消融
glacier spill-way 冰川隘口
glacier stadial 冰阶
glacier stage 冰阶段
glacier stage snowline 冰期雪线
glacier stage stratification 冰川地
　层法
glacier stagnation 冰川停顿
glacier staircase 冰台阶
glacier stairway 冰川阶梯
glacier stream 冰川河流
glacier striae 冰擦痕
glacier substage 冰期亚阶段
glacier summer resort 冰川避暑
　胜地
glacier surface velocity 冰川表面
　流速
glacier surge 冰川跃动
glacier table 冰桌
glacier temperature 冰川温度
glacier terminus (＝glacier terminal)
　冰川末端
glacier theory 冰川理论
glacier till 冰碛物
glacier tongue (＝glacier snout)
　冰舌
glacier tour 冰川旅游
glacier tourism 冰川旅游业
glacier tourism area 冰川旅游区

glacier tourism development 冰川
　旅游开发
glacier tourism planning 冰川旅
　游规划
glacier tourist 冰川旅游者
glacier tourist destination 冰川旅
　游目的地
glacier tourist equipment 冰川旅
　游装备
glacier tourist facilities 冰川旅游
　设施
glacier tourist infrastructure 冰川
　旅游基础设施
glacier tourist resources 冰川旅游
　资源
glacier transport 冰川搬运
glacier trekking (＝glacier hiking)
　冰川跋涉
glacier trip (＝glacier journey) 冰
　川旅游
glacier trough 冰川槽
glacier type 冰川类型
glacier uvala 冰川宽谷
glacier valley floor step 冰谷阶
glacier variation 冰川变化
glacier varve 冰川纹泥
glacier velocity 冰川运动速度
glacier vestige 冰川遗迹
glacier water 冰川水
glacier water resources 冰川水资源
glacier wedding 冰川婚礼
glacier well (＝moulin) 冰川竖井
glacier wind (＝ice wind) 冰川风

glacier zone 冰川作用区
glacieret 小冰川
glacifluvial 冰川洪积
glacigenic lithofacies 冰川相(岩石学,地层学)
glacigenous (=glacigenic) 冰成的,冰川的
glacioaqueous (=glaciofluvial) 冰水的
glacioaqueous clay 冰水泥
glaciofluvial (=glacioaqueous) 冰水的
glaciofluvial deposit 冰水沉积物
glaciofluvial deposition 冰水沉积作用
glaciofluvial environment 冰水环境
glaciofluvial erosion 冰水侵蚀
glaciofluvial fan 冰水扇
glaciofluvial landform 冰水地形
glaciofluvial stratigraphy 冰水地层学
glaciofluvial terrace 冰水阶地
glaciofluvial transport 冰水搬运
glacio-isostatic uplift 冰川均衡抬升
glaciokarst 冰川喀斯特
glaciokarst landscape 冰川喀斯特景观
glaciolacustrine 冰湖的
glaciolacustrine environment 冰湖环境
glaciolacustrine stratigraphy 冰湖地层学
glaciologist 冰川学家
glaciology 冰川学
glaciomarine environment 冰海环境
glaistels 厚层冰有机寒冻土
glare ice 光滑冰
glarosion 冰川侵蚀作用
glass ice 透明薄冰层
glaucous gull (*larus hyperboreus*) 北极鸥
glaucous ice 淡灰绿色冰
glaze 雨凇,冻雨
glaze ice 雨凇
glazed frost 雨凇,冻雨冰
gleization 潜育作用
gley horizon 潜育层
gleyic aqui-gelic cambosols 潜育潮湿寒冻雏形土
gleysolic static cryosol 潜育静态寒冻土
gleysolic turbic cryosol 潜育扰动寒冻土
glime 半透明冰,雨雾凇
glimmer ice 闪光冰
glitter 岩屑锥
global air pollution 全球大气污染
Global Atmosphere Watch (GAW) 全球大气本底站
global carbon cycle 全球碳循环
Global Carbon Project (GCP) 全球碳计划

global change 全球变化
Global Change and Terrestrial Ecosystems(GCTE) 全球变化与陆地生态系统
Global Change Observation Mission(GCOM) 全球变化观测计划
Global Earth Observing System of Systems(GEOSS) 全球综合地球观测系统
Global Energy and Water Cycle Experiment(GEWEX) 全球能水循环试验
global ice volume 全球冰量
Global Land Data Assimilation System(GLDAS) 全球陆地资料同化系统
global mean temperature 全球平均温度
global meteorological data 全球气象资料
Global Navigation Satellite System(GLONASS) 全球导航卫星系统
Global Observing System(GOS) 全球观测系统
Global Positioning System(GPS) 全球定位系统
global radiation 总辐射
global scale 全球尺度
global spectral model 全球谱模式
global telecommunication system(GTS) 全球电信系统(气象)
global transport 全球输送
global warming 全球变暖
global warming potential(GWP) 全球增温潜势
global water cycle 全球水循环
gamma monitor(GMON) 伽马射线雪水当量测量仪
gobi 戈壁
governing equation 控制方程
GPS aerotriangulation GPS空中三角测量
GPS receiver GPS接收机
graded river 均衡河流,坡度平缓的河流
gradient 梯度
gradient current 〔海洋〕梯度流
gradient observation 梯度观测
gradient wind 梯度风
grain 晶粒
grain boundary 晶粒边界
grain coarsening 颗粒粗化
grain orientation 晶粒取向
grain shape 晶型
grain size 晶粒尺寸,粒度
Grand maximum (中世纪太阳黑子)极大期
granic cryogenic fabric 粒状冷生结构
granular ice 粒状冰
granular snow 粒状雪
granular structure 粒状结构
granularity 粒度
granulometry 粒度测定
graphic data base 图形数据库

grass-bog peat　草本泥炭沼泽
grassland　草地
grating　光栅
graupel　霰，软雹
graveyard mounds　残留冰楔群
gravimeter　重力仪
gravimetric baseline　重力基线
gravimetric point　重力点
gravimetric water content　重量含水量
gravimetric (=gravity) station　重力测站
gravimetry　重力测量学，重量〔分析〕法
gravitational erosion　重力侵蚀
gravitational potential　引力位
gravitational stability　重力稳定度
gravitational tide　引力潮
gravitational wave　重力波
gravity anchor　重力锚
gravity anomaly　重力异常
gravity datum　重力基准
gravity field　重力场
gravity gradient measurement　重力梯度测量
gravity gradiometer　重力梯度仪
gravity measurement　重力测量
gravity potential　重力位势
gravity profiling　重力剖面探测方法
gravity survey　重力测量
gravity water　重力水
gravity wave　重力波

gravity wave drag parameterization　重力波拖曳参数化
gray absorber　灰色吸收体，灰体
gray atmosphere　灰体大气
gray body　灰体
gray ice　灰冰（海冰）
gray level　灰度值
gray plate　灰板
gray scale　灰阶
gray-level co-occurrence matrix (GLCM)　灰度共生矩阵
gray-white ice　灰白冰（海冰）
grazing　放牧
grazing angle　入射余角
grease ice　脂状冰
great interglacial　大间冰期
great pluvial　大雨期
Greater Cold　大寒（节气）
Greater Heat　大暑（节气）
green iceberg　绿冰山
greenhouse climate　温室气候
greenhouse effect　温室效应
greenhouse forcing　温室强迫〔作用〕
greenhouse gas capture　温室气体捕获
greenhouse gas emission　温室气体排放
greenhouse gas removal　温室气体移除〔量〕
greenhouse gas reservoir　温室气体储库
greenhouse gas sink　温室气体汇

greenhouse gas source 温室气体源
greenhouse gas stabilization 稳定温室气体浓度
greenhouse gas storage 温室气体储库
greenhouse gases（GHGs） 温室气体
greenhouse warming 温室增温效应
Greenland 格陵兰
Greenland anticyclone 格陵兰反气旋
Greenland current 格陵兰海流
Greenland high 格陵兰高压
Greenland Ice Sheet 格陵兰冰盖
Greenland Ice Sheet Project-2（GISP2） 格陵兰冰芯计划Ⅱ
Greenland shark (*somniosus microcephalus*) 格陵兰鲨鱼
Greenpeace 绿色和平组织
Greenwich Time（GT） 格林尼治时间
grey absorber 灰吸收体
grey body radiation 灰体辐射
grey ice 灰冰
grey level transform 灰度变换
grey white ice 灰白冰
grid bearing 坐标方位角
grid computing 网格计算
grid map 网格地图
grid mesh 格网
grid method 网格法
grid point 网格点,格点
grid point model 格点模式
grid resolution 网格分辨率
grid structure 网格结构
gross ablation 总消融
gross annual ablation 年总消融量
gross annual accumulation 年总积累量
gross fog 浓雾
gross photosynthetic rate 总光合作用速率
gross primary productivity（GPP） 总初级生产力
ground clutter 地物杂波
ground control point（GCP） 地面控制点
ground echo 地物回声,地面回波
ground fog 地面雾
ground freezing 土冻结
ground freezing period 土冻结期
ground ice 地下冰
ground ice mound 冰丘
ground inversion 地面逆温
ground moraine（＝bottom moraine） 底碛
ground nadir point 地底点
ground penetrating radar（GPR） 探地雷达
ground range resolution 地面距离分辨率
ground receiving radius 地面接收半径
ground receiving station 地面接收站
ground resolution 地面分辨率

ground settlement 地表下沉
ground station 地面站
ground subsidence 地表沉陷
ground temperature 地温
ground tilt measurement 地倾斜观测
ground track 地面轨迹
ground truth 地面真值,地面实况
ground veins 土脉
ground water deposit 地下水沉积物
ground water resources 地下水资源
ground water ŕunoff 地下径流
ground wedge 土楔
ground-based observation 地基观测
ground-based radar 地基雷达
ground-based Synthetic Aperture Radar（SAR） 地基合成孔径雷达
ground-based system 地基系统
grounded hummock 底冰丘,搁浅冰丘
grounded ice（＝stranded ice） 冻透冰(河、湖、海岸冰),搁浅冰
ground-ice mound 底冰丘
grounding line 接地线,落地线,触地线
groundwater 地下水
groundwater artery 地下水干道
groundwater basin 地下水盆地
groundwater contour 地下水位等值线
groundwater dating 地下水定年
groundwater decrement 地下水消耗
groundwater depletion 地下水枯竭
groundwater divide 地下水分水岭
groundwater elevation 地下水位
groundwater feed 地下水补给
groundwater flow 地下水〔径流〕,基流
growing season 生长季节
growler 碎浮冰(海冰)
growth curve 增长曲线
growth ring 年轮,生长轮
guidebook 旅游指南
guillemot 海鸠
Gunz Glacial Stage 恭兹冰期
Gunz-Mindel Interglacial Stage 恭兹一民德间冰期
gymnosperm 裸子植物
gypsic anhyorthels 石膏脱水正常寒冻土
gypsic anhyturbels 石膏脱水扰动寒冻土
gyro azimuth 陀螺方位角
gyrophic EDM traverse 陀螺定向光电测距导线
gyrostatic orientation survey 陀螺仪定向测量

H

habitat 生境,栖息地
habitat loss 栖息地丧失
hada(＝kha-btags) 哈达
Hadley cell 哈得来环流〔圈〕
Hadley regime 哈得来域,哈得来对称流型(转盘实验的术语)
hail 冰雹
hail embryo 雹胚,雹核,冰雹胚胎
hail growth 冰雹增长
hail mitigation 消雹
hail pellet 雹粒
hailfall 降雹
hair hygrograph 毛发湿度计
hair hygrometer 毛发湿度表
half-life (放射性元素)半衰期
halo 晕,光晕
halocarbon 卤化碳
halocline 盐跃层
halogen 卤素
halogen acid 氢卤酸
hammock 冰胀丘
hand anemometer 手持风速表
hanging alluvial fan 悬冲积扇
hanging cirque 悬冰斗
hanging dam 悬冰坝(河流)
hanging fjord 悬式峡湾
hanging glacier 悬冰川
hanging tributary valley 悬支谷
hanging valley 悬谷
haplorthels 弱育正常寒冻土
haploturbels 弱育扰动寒冻土
hard freeze 深度冻结
hard frost 黑霜
hard rime 霜凇
hard windpack snow (雪面或雪层内)风板
hardness 硬度
hardy crop 耐寒作物
hardy plant 耐寒植物
hardy variety 耐寒种
harmonic Synthetic Aperture Radar (SAR) 合成孔径谐波雷达
haze 霾
haze aerosol 霾气溶胶
haze aloft 高空霾
haze droplet 霾滴
headed esker 串珠状蛇形丘
headwall 后壁
headwater 河源,水源
heat balance 热平衡
heat budget 热平衡,热收支
heat capacity 热容量
Heat Capacity Mapping Mission (HCMM) 热容量成像卫星
heat conduction 热传导
heat conductivity 导热率,热导率
heat content 热含量
heat diffusion 热扩散

heat energy 热能
heat exchange coefficients 热交换系数
heat flow 热流
heat flow plate 热流板,热流计
heat flux 热通量
heat flux budget 热通量收支
heat flux vector 热通量矢量
heat island effect 热岛效应
heat pipe 热桩,热管
heat pump 热泵
heat radiation 热辐射
heat recovery devices 热回收装置
heat resources 热量资源
heat sink 热汇
heat source 热源
heat storage 热储量
heat transfer equation 传热方程
heating index 采暖指数
heating rate 加热率,增温率
heating structure 采暖建筑物
heaving 冻胀,隆起
heavy floe 厚浮冰块,厚凌
heavy ice 重冰(海冰)
heavy ice date 重冰日
heavy ice period 重冰时段
heavy metal 重金属
heavy rain 大雨
heavy snow 大雪
heavy snow warning 大雪预警
hekistotherm (=hekistothermic plant) 适寒植物
Hellmann recording snow gauge 赫尔曼自记雪量器
hemic glacistels 半腐厚冰层有机寒冻土
hemic soil materials 半腐有机土壤物质
hemispherical albedo 半球反照率
hemistels 半腐有机寒冻土
Hercules Dome 大力神冰穹(南极)
herdsman 牧民
hermic Fibri-Permagelic Histosols 半腐纤维多年冻结有机土
heterogeneous 非均匀的,多相的,异质的
heterogeneous atmosphere reaction 多相大气反应
heterogeneous flow 非均质流
heterogeneous medium 不均匀介质
heterogeneous reaction 非均相反应
heterogeneous system 非均相体系
hexagonal aeolotropy 六角体各向异性
hexagonal column 六角柱体
hexagonal platelet 六角板状
hexagonal system 六方晶系
hibernal 冬天的,冬令
hibernation 冬眠,越冬
high Asia 高亚洲
high Asia glacier 高亚洲冰川
high fill foundation 高填土地基
high frequency radar 高频雷达
High Resolution Imaging Spectrometer(HIRIS) 高分辨率成像光

谱仪
high-altitude wetlands （HAWs） 高海拔湿地
high-center polygon 凸心多边形土
highland climate 高地气候
highland field （= highland ice = highland glacier） 冰原
highly frost susceptibility 高冻胀敏感性
high-pass filter 高通滤波器
high-polar glacier 高纬极地冰川
high-strength ice 高强度冰
high-voltage pulse 高压脉冲
highway frost boiling 公路翻浆
Himalayan Mountains 喜马拉雅山脉
Himalayan Griffon（*gyps himalayensis*） 喜马拉雅狮鹫
Himalayan Snowcocks（*tetraogallus himalayensis*） 暗腹雪鸡
Himalayan Yew（*taxus wallichiana*） 喜马拉雅红豆杉
Himalayas 喜马拉雅山脉
Hindukush-Karakoram-Himalayas （HKH） 兴都库什－喀喇昆仑－喜马拉雅
histels 有机寒冻土
histic Aqui-Gelic Cambosols 有机潮湿寒冻雏形土
histic Dystric-Turbic Cryosol 有机不饱和扰动寒冻土
histic epipedon 有机表层
histic evidence 有机现象
histic Hapli-Permagelic Gleyosols 有机简育多年冻结纤维土
histic Molli-Gelic Cambosols 有机暗沃寒冻雏形土
histic Permi-Gelic Cambosols 有机多年冻结寒冻雏形土
histic Umbri-Gelic Cambosols 有机暗瘠寒冻雏形土
historical climate series 历史气候序列
historical climatic data 历史气候资料
historical climatology 历史气候学
historical geology 历史地质学
historthels 有机正常寒冻土
histoturbels 有机扰动寒冻土
hoar（= hoar frost） 白霜
hoar crystal 白霜〔冰〕晶
hoar frost（= hoar） 白霜
hodograph 高空风分析图
Holocene 全新世、全新统，全新世的、全新统的
Holocene climatic optimum 全新世气候适宜期，全新世最暖期
Holocene Epoch 全新世
Holocene Megathermal 全新世大暖期
Holocene permafrost 全新世多年冻土
Holocene Series 全新统
hologrammetry 全息摄影测量
Holographic Ice Surveying System （HISS） 全息冰层探测系统

holospheric temperature　全球温度,纬圈平均温度
holosteric barometer　固体气压表,空盒气压表
hommock　冰丘
homogeneous flow　均质流
homogeneous medium　均匀介质
homogeneous porous medium　均匀多孔介质
homosphere　均质层(大气)
horizon　层位,地平线
horizon track　层位踪迹
horizontal axis　水平轴
horizontal change　水平变化
horizontal coherence　水平相干性
horizontal component　水平分量
horizontal compression　水平压缩
horizontal control network　平面控制网
horizontal control point　平面控制点
horizontal coordinate　平面坐标
horizontal cross-section　水平剖面
horizontal divergence　水平散度
horizontal expansion　水平延伸
horizontal flow　水平流动
horizontal frost-heave force　水平冻胀力
horizontal gradient　水平梯度
horizontal ice force　水平冰力
horizontal movement　水平移动
horizontal normal fold　水平正常褶皱
horizontal overlap　水平掩覆
horizontal resolution　水平分辨率
horizontal scale　水平尺度
horizontal strain　水平应变
horizontal stratification　水平层理
horizontal stress　水平应力
horizontal zonality　水平地带性
horizontal-dipole sounding　水平偶极测深
horn　角峰
Hot-water drill　热水钻
Hui Nationality　回族
human poverty index (HPI)　人类贫困指数
humic acid　腐殖酸
humic epipedon　腐殖质表层(米)
humic organic cryosol　腐殖有机寒冻土
humic substance　腐殖物质
humid mesothermal climate　湿温气候
humid microthermal climate　湿润低温气候(雪林气候)
humidification　湿化
humification　腐殖化作用
hummock　波状地,草皮小丘,冰丘,泥炭丘
hummock ice　冰丘冰
hummock-and-hollow topography　鼓盆地形
hummocking　浮冰集群,浮冰拥塞
hummocky floe　堆积浮冰块,冰丘冰原,冰丘浮块
hummocky microrelief　泥丘微地形

hummocky moraine 冰碛丘陵
hummocky terrain 丘陵地形
hummocky topography 丘陵地形
humpback whale 座头鲸
humus 腐殖质
humus horizon 腐殖质层
hyaloclastite 玻质碎屑岩，玄武碎屑岩，碎玻质熔岩
hydrate 水合物
hydraulic conductivity 水力传导率，水力传导系数
hydraulic depth 水压深度
hydraulic diffusivity 水力扩散系数
hydraulic permeability 水力渗透性
hydraulic property 水力学性质
hydraulic thawing 水力融化
hydrocarbon 碳氢化合物
hydrochemical talik 水化学融区
hydrochloric acid 盐酸，氢氯酸
hydrochlorofluorocarbons（HCFCs） 氢氟氯碳化物
hydrodynamic instability 流体动力不稳定〔性〕
Hydrofluorocarbons（HFCs） 氢氟碳化物
hydrogel 水凝胶
hydrogen 氢
hydrogen chloride 氯化氢
hydrogen ion 氢离子
hydrogen ion concentration（pH） 氢离子浓度
hydrogen peroxide 过氧化氢，双氧水
hydrogen sulfide 硫化氢
hydrogeologic section 水文地质剖面
hydrogeologic survey 水文地质调查
hydrogeological environment 水文地质环境
hydrogeothermal resources 地热资源
hydrographic chart 水文图
hydrographic survey 水文测量
hydrography 水文地理学
hydrologic cycle（＝hydrological cycle） 水文循环
hydrologic effect 水文效应
hydrologic forecast 水文预报
hydrologic gradient 水文梯度
hydrologic map 水文图
hydrologic modeling 水文建模，水文模拟
hydrologic simulation 水文模拟
hydrological cycle（＝hydrologic cycle） 水文循环
hydrological mark 水文标记
hydrological store 水分储存
hydrological stratigraphy 水文地层学
hydrologist 水文工作者，水文学家
hydrology 水文学
hydrometeor 水凝物，水凝现象
hydrometeorological forecast 水文气象预报
hydrometeorology 水文气象学
hydrometric method 水文测定法
hydrometry 水文观测
hydrophilic 亲水的，亲水性

hydrophobic 疏水的,疏水性
hydrophysical model 水文物理模型
hydrophysics 水文物理学
hydroscopic nuclei 吸湿性核
hydrosol 水溶胶
hydrosphere 水圈
hydrostatic approximation 静水近似
hydrothermal glaciology 水热冰川学
hydrothermal talik 水热融区
hydroxyl 羟基
hydroxyl radical（OH） 羟基,氢氧基
hygrogram 湿度自记曲线
hygrograph 湿度计
hygrometer 湿度表
hygronics 电子湿度计
hygronom 湿度仪
hygroscopic 吸湿的
hygroscopic moisture 吸湿含水量
hygroscopic nuclei 吸湿性核
hygroscopic particle 吸湿性粒子
hygroscopicity 吸湿性
hygrostat 恒湿器,湿度检定箱
hygrothermogram 温湿自记曲线
hygrothermograph 温湿计
hygrothermometer 温湿表
hygrothermoscope 温湿仪
hygrothermostat 恒温恒湿器
hyperborean 极北寒冷区
Hyperion 地球观测卫星-1（EO-1）上携带的高光谱成像仪（美国）
hyperspectral image 高光谱影像
Hyperspectral Imaging Sensor (HIS) 超光谱成像仪
hyperspectral remote sensing 高光谱遥感
hyperthermal stage 高温期
hypsithermal 高温期,气候最适宜期
hypsithermal interval 高温时段（冰期后）
hysteresis 滞后现象,滞后作用
hythergraph 温湿图

I

ice abrasion 冰磨蚀
ice accretion 积冰
ice age（＝glacial age） 冰期
ice age aridity 冰期干燥性
ice age cycle 冰期旋迴
ice albedo 冰反照率
ice albedo feedback 冰反照率反馈〔效应〕
ice algea 冰藻
ice analysis 冰情分析

Ice and Snow Festival 冰雪节	ice break 冰裂,解冻,开冻,开河
ice apparatus 测冰器	ice breaker 破冰船,破冰器
ice appearance forecast 初冰预报	ice breaking service 破冰作业
ice apron 冰裙	ice breccia (= ice mosaic) 混冻冰,角砾冰
ice arena 滑冰场	
ice atlas 冰图集	ice bridge 冰桥
ice auger 凿冰器,冰钻	ice bridge effect 冰桥效应,冰塞效应
ice avalanche 冰崩	
ice ax 冰镐	ice buckling 冰崩,冰解
ice axe 冰斧	ice bulb temperature 冰球温度
ice ballet 冰上芭蕾	ice cache 冰窟
ice bank 冰岸,冰带	ice caisson 冰沉箱
ice barchan 新月形冰丘	ice cake 冰饼,冰块
ice barrier 冰障,冰堵,冰壁	ice calorimeter 冰热量计,冰卡表
ice barrier basin 冰围盆地	ice calving 裂冰,冰崩解
ice basket 冰网	ice canopy 冰顶盖,冰冠(海冰)
ice bay (= ice bight) 冰凹湾	ice cap 冰帽
ice bed (rock) interface 冰－岩界面	ice carapace 冰被
ice belt 浮冰带,冰带	ice carve 冰雕
ice bending 冰面凸凹(海冰)	ice cascade 冰瀑布
ice berg 冰山	ice cast 冰铸型
ice bight (= ice bay) 冰凹湾	ice cave 冰穴,冰洞
ice blade 冰刀	ice cavern 冰洞
ice blink 冰反光区,冰映光	ice chart 冰情图,冰区图,海冰图
ice blister 冰丘	ice chisel 冰凿子
ice block ridge 冰块脊	ice chute 泄冰道
ice body 冰体,冰块	ice class 冰级
ice boom 防冻栅,冰轰声	ice clearing 冰间湖
ice borehole 冰钻孔	ice cliff 冰崖
ice boulder 冰川巨砾	ice climate 冰雪气候
ice bound 冰封区,冰封的,被冰堵塞的	ice clogging 冰塞
	ice cloud 冰晶云
ice boundary 冰界	ice cluster 大冰堆

ice coating 结冰,冰层
ice code 冰情符号,冰情电码
ice collapse 冰塌,冰崩
ice column 冰柱
ice concentration 覆冰量,海冰密集度
ice condition 冰情
ice condition forecast 冰情预报
ice condition statistics 冰情统计
ice condition summary 冰情概要,冰况概要
ice cone 冰锥
ice contact feature 冰界特征
ice content 含冰量
ice control 冰控制
ice control structure 冰流控制建筑物,防冰建筑物
ice core 冰芯
ice core dating 冰芯定年
ice core profile 冰芯剖面
ice core record 冰芯记录
ice cover 覆冰(河冰,湖冰)
ice coverage 冰覆盖率
ice covered channel 冰覆盖水道
ice cracking engine 破冰机
ice creep 冰蠕变,冰蠕动
ice crevasse 冰裂隙
ice crossing 冰桥
ice crust 冰皮,冰壳,薄冰壳
ice crystal 冰晶
ice crystal cloud 冰晶云
ice crystal concentration 冰晶浓度
ice crystal effect 冰晶效应
ice crystal fabrics 冰晶组构
ice crystal fog 冰晶雾
ice crystal form 冰晶状
ice crystal growth 冰晶增长,冰晶生长
ice crystal haze 冰晶霾
ice crystal modification 冰晶变形
ice crystal nucleus 冰晶核
ice crystal orientation 冰晶方位,冰晶取向(C轴取向)
ice crystal profile 冰晶体剖面
ice crystal size 冰晶体尺寸
ice crystal structure 冰晶〔体〕结构
ice crystallography 冰结晶学
ice current trace 冰溜遗痕,冰擦痕
ice cutter 冰钻头,切冰刃
ice dam 冰坝
ice dam burst 冰坝溃决
ice damage 冰情灾害,冰灾损失
ice dammed lake 冰坝湖
ice day 有冰日
ice debris layer 冰屑层
ice deformation 冰体形变,冰体变形
ice density 冰体密度
ice deposit 积冰
ice detector 测冰仪
ice dike 冰脉
ice discharge 冰流量
ice disintegration 冰崩解作用
ice divide 分冰岭

ice dome   冰穹
ice drift（＝ice run＝ice floe＝ice float）   流冰,漂冰,淌凌,冰流
ice drill   冰钻
ice duration   结冰期
ice dynamics   冰动力学
ice edge   冰区边界,冰缘线
ice effect   冰效应
ice engineering   冰工程
ice environment   冰环境
ice eroded ripple   冰蚀波痕
ice erosion   冰蚀
ice extent   海冰范围
ice fabrics   冰组构
ice fall   冰瀑布,冰崩
ice feather   冰羽,霜羽,冰凇
ice fender   冰区护舷材
ice field   冰原
ice field belt   冰原带
ice float   浮冰,冰盘
ice floe   浮冰块
ice floe velocity   浮冰流速,冰块流速
ice flood   凌汛
ice flow   冰流动,冰流
ice flow dynamics   冰流动力学
ice flow model   冰流模型
ice flower   冰花
ice flowline   冰流线
ice flux   冰通量
ice fog   冰雾
ice foot   冰脚,冰壁,冰栅
ice force   冰作用力

ice force calculation   冰力计算
ice forecast   冰情预报
ice formation   结冰,成冰,成冰作用
ice formation age   成冰年龄
ice formation condition   结冰条件
ice fragmenting device   破冰装置
ice free   无冰
ice free area   无冰区
ice free period   无冰期
ice free port   不冻港
ice friction   冰摩擦
ice fringe   冰缘
ice front   冰锋面,冰崖,冰线冰崖
ice gang   流冰,融冰流,淌凌
ice gate   排冰闸门,泄冰闸门
ice gauge   量冰尺
ice gland   (雪里的)冰脉
ice gland serac   冰柱
ice gorge   冰坝,冰谷,冰峡
ice grain   冰晶粒
ice granule   冰簌
ice growth rate   冰增长率,冰生长速率
ice gruel   冰泥
ice guard   挡冰栅,冰挡,破冰设备
ice gush   冰涌
ice hammer   冰锤
ice hanging dam   悬冰坝
ice harbor   冻港
ice heap   冰堆
ice hillock   (小)冰丘,(小)冰堆
ice hockey   冰球(运动)

Ice Hockey Tournament 冰球锦标赛
ice hummock 冰堆积,冰丘
ice hydrometeor 结冰物,冰冻物
ice impact pressure 流冰冲击压力
ice index 冰封指数
ice inhomogeneity 冰体非均匀性
ice island 浮冰岛
ice isthmus 冰峡
ice jam 冰塞,冰壅塞,冰障
ice jam characteristic 冰壅特征
ice jam flood 冰塞凌汛,流冰凌汛,凌汛
ice keel 冰龙骨
ice kinematic parameter 冰运动学参数
ice lamina 薄层冰
ice lane 冰道
ice lantern 冰灯
ice layer 冰层
ice lead 冰间水道,水沟
ice ledge 冰瀑,冰壁,冰栅,沿岸冰
ice lens 冰透镜体
ice limit 冰区边界,冰限
ice line (冰水平衡图上的)冰线
ice load 冰载荷
ice lobe 冰垂,冰舌,冰川舌
ice margin 冰边缘
ice marginal lake 冰川边缘湖
ice marginal stream 冰川边缘河
ice mark 冰痕
ice mass 冰块
ice mass flux 冰通量
ice mechanics 冰力学
ice mill 冰川砾磨蚀地,冰砾磨蚀地
ice mixing ratio 冰水混合比
ice morphology 冰形态学
ice mosaic (= ice breccia) 混冻冰,角砾冰
ice motion 冰体运动
ice motion algorithm 冰体运动算法(海冰)
ice mound 冰丘
ice multiplication 冰粒增多
ice mushroom (= ice table = ice pedestal) 冰蘑菇,冰桌
ice navigation 冰区航行
ice needle 冰针
ice nil 无冰
ice nucleus (pl. nuclei) 冰核
ice observation 冰情观测
ice pack 堆积冰,积冰
ice pan 冰盘
ice particle 冰粒
ice particle concentration 冰粒密集度
ice particle counter 冰粒计数器
ice pass 泄冰建筑物(泄冰孔,泄冰闸门等)
ice patch 流冰
ice patrol 冰情巡逻船
ice patrol service 冰情巡逻站,浮冰观测站
ice pavement 冰坪

ice pedestal (= glacier table = ice table = ice mushroom) 冰蘑菇,冰桌
ice pellet 冰丸,小冰球,小冰雹,冰粒,冰珠
ice penetrating radar 测冰雷达
ice penitentes (= ice pinnacle) 冰塔林
ice period 结冰期(河、湖、海冰)
ice petrology 冰岩学
ice phase 冰相
ice phase parameterization 冰相参数化
ice pillar 冰柱,霜柱
ice pinnacle (= ice penitentes) 冰塔林
ice platform 冰质平台
ice platform drilling 冰质平台钻井,冰基钻井
ice point 冰点
ice pole 冰极(约位于北纬84°,西经160°)
ice port 冰码头
ice prediction technique 冰况预测技术
ice pressure 冰压力
ice prevention 防凌,防冰
ice prevention measures 防冰措施
ice prism 冰晶柱,冰针
ice profile sonar 测冰声呐
ice properties 冰特性
ice push 冰壅
ice pyramid 冰塔

ice quake 冰震
ice radar 冰雷达
ice raft 浮冰,冰筏
ice rafting 冰筏作用
ice rain 冻雨
ice ramp 冰坡
ice rampart (河冰、湖冰推挤作用形成的)岸堤
ice receiving area 冰补给区
ice reconnaissance 冰情侦察
ice regime 冰情,冰况
ice regime chart 冰情图
ice regime code 冰情符号
ice regime feature 冰情特征
ice regime forecast 冰情预报
ice regime observation 冰情观测
ice region 冰区
ice report 冰情报告
ice reservoir area 冰库区
ice resistant semisubmersible drilling unit (IRSDU) 抗冰半潜式平台/装置
ice rheology 冰流变学
ice ribbon 冰条纹,冰条带
ice ridge 冰脊
ice ridge thickness 冰脊厚度
ice rind 冰壳
ice rise 冰隆,冰丘
ice rock avalanche 冰岩雪崩
ice rock mixture 冰岩混合体
ice routing 冰区航线
ice ruler 冰尺
ice rumple 冰褶皱

ice run 冰流,冰汛	ice skate blade 冰刀
ice run concentration 流冰疏密度	ice skating 滑冰
ice sample 冰样	ice skin 冰衣,冰皮,薄冰壳
ice saturation 冰饱和	ice sky 冰照云光
ice scape 冰景(特指极地景观)	ice skylight 冰天窗(从潜艇中观看)
ice scarp 冰蚀	
ice scouring 冰川挖掘作用	ice slab 厚冰层,冰板
ice screw 冰螺栓,岩钉	ice slid mark 冰成滑痕
ice season 结冰季节	ice slide 冰体滑动
ice section 薄冰切片	ice slope 冰坡
ice segregation 冰分凝作用	ice sludge 海绵冰,细碎浮冰
ice segregation ratio 冰分凝比率	ice sluice (=sluiceway) 泄冰道,泄冰闸
ice segregation theory 冰分凝理论	
ice shear plane 剪切冰面	ice slush 雪浆
ice sheet 冰盖	ice sounding radar 探冰雷达
ice sheet base 冰盖底部	ice spicule 冰针
ice sheet collapse 冰盖崩解	ice splinter 屑冰,碎冰片,碎冰片
ice sheet crest 冰盖穹顶	ice split 冰劈
ice sheet depression 冰盖低地	ice sports 冰上运动
ice sheet discharge 冰盖排出量	ice stagnation 冰滞期
ice sheet disintegration 冰盖分裂	ice stake (=ski stock) 雪杖
ice sheet dynamics 冰盖动力学	ice stalagmite 冰笋
ice sheet front 冰盖前缘	ice station 冰情观测站
ice sheet growth 冰盖生长	ice status 冰情
ice shelf 冰架	ice storm 暴风雪
ice shelf cavity 冰架海腔	ice stratigraphy 冰层理
ice shelf disintegration process 冰架崩解过程	ice stream 冰流
	ice stream catchment 冰流盆地
ice shelf disruption 冰架分解	ice strength 冰强度
ice shelf edge 冰架边缘	ice strip 窄冰带,浮冰带
ice shelf system 冰架系统	ice structure 冰结构,冰构造
ice shelf water 冰架水	ice surface feature 冰面特征
ice situation 冰情	ice surface lake 冰面湖

ice surface runoff　冰面径流
ice surge　（冰盖）快速扩张，冰川跃动
ice table（＝ice mushroom＝ice pedestal）　冰桌
ice temperature　冰温
ice texture　冰结构
ice thickness　冰厚度
ice thrust　冰川推挤，冰压力，冰推力
ice thrust ridge　（河、湖）冰推脊
ice tongue　冰舌
ice trench　（海冰）冰沟
ice tunnel　冰内通道，冰隧道
ice undercurrent　冰潜流
ice vein　冰脉
ice velocity　冰流动速度，冰运动速度
ice viscosity　冰黏滞性
ice volume　冰体积，冰量
ice wall　冰墙，冰壁，冰栅
ice warning　冰情预警
ice water content　（云内）冰水含量
ice water interface　冰水界面
ice water path　冰水通道
ice water slurry（＝ice water slush）　冰水浆，雪浆
ice wave　冰波
ice wedge　冰楔
ice wedge cast　冰楔假型
ice wedge complex　冰楔复合体
ice wedge polygon　冰楔多边形
ice wedge trough　冰楔槽

ice wind（＝glacier wind）　冰川风
ice worm　冰川蚯蚓
ice yacht　冰上帆橇
ice zone　冰区
Ice，Cloud，and Land Elevation Satellite（ICESat）　冰、云和陆地高程卫星
ice-accretion indicator　积冰指示器
ice-bearing current　冰凌，挟冰水流
ice-bearing permafrost　含冰多年冻土
iceberg calving　冰山崩解
iceberg deposit　冰山沉积
iceblink　冰反光
ice-blocked lake　冰塞湖
ice-bonded permafrost　冰胶结多年冻土
ice-border drainage　冰川边缘水系
ice-boring machine　冰钻机
ice-borne sediment　冰积物
ice-bottom reflection　冰底反射
ice-bound season　封冻期
ice-breaking capability　破冰能力
ice-breaking carrier　破冰船
ice-cemented　冰胶结的
ice-cemented debris　冰胶结岩屑
ice-covered area　冰覆盖区
ice-crystal mark　冰晶痕
ice-crystal theory　冰结晶学（理论）
ice-dynamic instability　冰动力不稳定性
ice-equivalent thickness　冰当量厚度
ice-free cirque　空冰斗

ice-free season 无冰季节,无冰期
ice-free waterway 无冰水道
ice-free zone 无冰区
ice-jam prevention 防凌
ice-jam stage 壅冰水位
Icelandic low 冰岛低压
ice-medium permafrost 中等含冰量多年冻土
ice-nucleation temperature 结冰温度
ice-ocean-atmosphere interactions 冰-海-气相互作用
ice-poor permafrost 少冰多年冻土
ice-pressed moraine 挤压冰碛
ice-push moraine 推出碛
ice-rafted boulder 浮冰搬运漂砾
ice-rafted debris 冰筏岩屑(沉积)
ice-reinforced drillship 抗冰钻井船
ice-rich permafrost 富冰多年冻土
ice-rich zone 富冰带
ice-saturated permafrost 饱冰多年冻土
ice-scour lake 冰蚀湖
ice-scoured plain 冰蚀平原
ice-sculptured pyramid 冰蚀塔
iceworn 冰蚀的,冰损的
icicle 冰柱子,冰溜子
iciness 结冰程度
icing 冰锥,积冰
icing blister 冰锥复合体
icing glade 冰锥间空地
icing ice 冰锥冰

icing index 结冰指数
icing intensity 积冰强度
icing level (大气层)积冰高度
icing meter 积冰表
icing mound 冰锥丘
icing-rate meter 积冰速率计
icy firn 冰结粒雪
icy permafrost 富冰多年冻土
Ikonos satellite 伊科诺斯卫星(美国)
illustrative map units 图解制图单元
image 影像
image analysis 图像分析
image classification 图像分类
image coding 图像编码
image correlation 图像相关
image data compression 图像压缩
image database 影像数据库
image enhancement 图像增强
image fusion 影像融合
image histogram 图像直方图
image matching 影像匹配
image mosaic 图像镶嵌
image overlaying 图像叠加
image processing 图像处理
image projection transformation 图像投影变换
image pyramid 影像金字塔〔显示〕
image quality 图像质量
image recognition 图像识别
image reprojection 图像重投影
image resolution 影像分辨率

image segmentation 图像分割
image transformation 图像变换
imager 成像仪
imaging radar 成像雷达
imaging spectrometer 成像光谱仪
immediate runoff 直接径流
immigrant 移入者
impermeability 抗渗性
impermeable layer 不透水层
impervious area runoff 不透水区径流
impervious formation 不透水层
imperviousness 不透水性
Improved TIROS Operational Satellite (ITOS) 改进型"泰罗斯"业务卫星
impure ice 含杂质冰
impurity concentration 杂质浓度
impurity content 杂质含量
impurity element 杂质元素
in situ measurement 实地测量
in situ monitoring 实地监测
in situ observation 实地观测
in situ processing 现场处理
inactive ice wedge 不活动冰楔
inactive moraine 不活动冰碛
inactive rock glacier 不活动石冰川
incidence angle 入射角
incidence plane 入射面
incident energy 入射能量
incident wave 入射波
incipient crack 初始裂纹
incipient fracture 初始断裂

inclination angle 倾角
incoherence 非相干,松散性
incoming radiation 入射辐射
incompatibility 不相容性
incompressibility 不可压缩性
incremental scenarios 增量情景
Indian monsoon 印度季风
Indian Remote Sensing Satellite (IRS) 印度遥感卫星
indicator community 指示群落
indigenous culture 本土文化
inert gas 惰性气体
inertia stress 惯性力
Inertial Navigation System (INS) 惯性导航系统
Inertial Surveying System (ISS) 惯性测量系统
inextensibility 不可拉伸性
infiltration 入渗,渗侵,渗透
infiltration capability 渗透能力
infiltration capacity 渗透容量
infiltration capacity curve 渗透容量曲线
infiltration ice 渗浸冰
infiltration index 渗透指数
infiltration rate 渗透速率
inflexion point 拐点
inflow 入流,补给量
inflow boundary 入流边界
inflow curve 入流量曲线
inflow discharge 入流流量
inflow volume 入流量
influx 流入,汇流

information access 信息存取
information acquisition 信息获取
information attribution 信息属性
information extraction 信息提取
information standardization 信息标准化
infraglacial accumulation 冰底堆积过程
infraglacial deposit 冰下沉积
infrared absorption spectroscopy 红外吸收光谱学
infrared image 红外影像
infrared imagery 红外图像
infrared photography 红外摄影
infrared radiation 红外辐射
infrared radiometer 红外辐射计
infrared remote sensing 红外遥感
infrared scanner 红外扫描仪
infrared spectrograph (IRS) 红外多光谱测量
infrared-stimulated luminescence (IRSL) dating 红外释光测年
inherent frequency 固有频率
inherent mode 固有模态
inherent stability 固有稳定性
inherent stress 内在应力
inhomogeneity 非均匀性
inhomogeneous flow 非均匀流
inhomogeneous medium 非均匀介质
initial border ice 初生岸冰
initial boundary condition 初始边界条件
initial condition 初始条件
initial freezing 初始冻结
initial ice 初生冰
initial water content of frost-heaving 起始冻胀含水量
injection ice 侵入冰
inland 内陆
inner boundary layer 内边界层
inner friction coefficient 内摩擦系数
inner friction force 内摩擦力
inner moraine 冰内碛
inorganic carbon 无机碳
insolation 日射,日照
insoluble 不溶的
insoluble particles 不溶微粒
instability 不稳定性
instability constant 不稳定常数
instability criterion 失稳准则
instability mode 不稳定状态
instability threshold 不稳定性阈值
instantaneous change 瞬时变化
instantaneous field of view (IFOV) 瞬时视场
instantaneous ice impact 瞬时冰冲击
instrumental analysis 仪器分析
instrumental error 仪器误差
insulation thickness 保温层厚度
integrated assessment 集成评估
integrated database 综合数据库
Integrated Global Water Cycle

Observation (IGWCO)　全球综合水循环观测
integrated pest management (IPM)　有害生物综合治理
integrated positioning　组合定位
integrated services digital network (ISDN)　数字通信网络
integration time　积分时间
intensity ratio　密集度比
intensive observation　加强观测,密集观测
interannual climate variability　年际气候变率
interannual time scale　年际尺度
interannual variability　年际变率
interception　截留,截留量
intercontinental sea　陆间海
intercrystalline cracking　晶间裂隙
intercrystalline fracture　晶间破裂
intercrystalline rupture　晶间破坏
interdecadal　年代际
interdiluvial period　间洪积期
interfacial effect　界面效应
interfacial tension force　界面张力
interfacial velocity　界面速度
interfacial viscosity ratio　界面黏度比
interfacial water　界面水
interfacial wave　层面波
interfere　干涉
interference effect　干涉效应
interferometric spectrometer　干涉分光计
Interferometric Synthetic Aperture Radar (InSAR)　干涉合成孔径雷达
interglacial　间冰期〔的〕
interglacial age (= interglacial)　间冰期
interglacial epoch　间冰期
interglacial lake　冰川间湖
interglacial period (= interglacial phase = interglacial stage = interglacial epoch = interglacial age)　间冰期
interglacial stadial　间冰段
interglacial stage　间冰期
intergranular　晶粒间的,颗粒间的
intergranular pressure　晶粒间压力,颗粒间压力
intergranular stress　粒间应力
interior drainage　内陆水系
interior lake　内陆湖,内流湖
interior layer　内部层
intermediate discontinuous permafrost　中等不连续多年冻土
intermediate moraine　中碛垄
intermittent ice lens　不连续冰透镜体
intermont basin (= intermontane basin)　山间盆地
intermontane basin (= intermont basin)　山间盆地
intermontane trough　山间槽地,岛弧区海槽
intermorainal lake　碛内湖

internal friction　内摩擦力
internal glacio-stratigraphy　冰川内部层理
internal ice deformation　冰内形变
internal ice layer　内部冰层
internal layer　内部层
internal radio-echo layer　内部回波层
internal reflection horizon（IRH）内部反射层
internal structure　内部结构
International Atomic Time（IAT）国际原子时
International Celestial Reference Frame（ICRF）　国际天球参考框架
International Geophysical Year（IGY）国际地球物理年
International Hydrography Organization（IHO）　国际海道测量组织
International Ice Patrol（IIP）　国际冰情巡逻队
international snow classification　国际雪形分类
International Terrestrial Reference Frame（ITRF）　国际地球参考框架
interpermafrost water　多年冻土层间水
interpolation error　插值误差
interstitial ice　空隙冰
interstitial water　隙间水,吸附水,束缚水
intertropical convergence zone（ITCZ）热带辐合区(带)
intine　（孢粉）内壁
intracrystalline　晶体内的
intraformational　层内的
intraformational contortion　层内扭曲
intraformational corrugation　层内揉褶
intraformational folding　层内褶皱
intraformational recumbent fold　层内伏卧褶皱
intrusive ice　侵入冰
Inuit　因纽特人,(北极)爱斯基摩人
inverse plummet observation　倒锤(线)观测
inverse problem　逆问题
inversion base　逆温层底
inversion height　逆温层高度
inversion layer　逆温层
involution　冰卷泥,冻融褶皱
inwash　冰川边缘沉积
ion bond　离子键
ion exchange　离子交换
ion origin　离子来源
ionic balance　离子平衡
ionic flux　离子通量
irradiance　辐照度
irradiation　辐照
irreversible process　不可逆过程
island permafrost　岛状多年冻土

isoband cryogenic fabric　微层状冷生微结构
isobar　等压线
isobaric surface　等压面
isobath　等深线
isochion　雪深等值线,雪日等值线
isochron dating　等时线定年
isogeotherm　等地温线
isogradient　等梯度线
isohyetal method　等雨量线法,等值线法
isolated cryopeg　冻土层内过冷盐土(液)
isolated permafrost　零星多年冻土
isolated talik　(多年冻土内)孤立融区
isonival　等雪量的
isopag(＝isopague)　等冻期线
isopectric　冰冻等时线
isopiestic line　等(水)压线
isopycnic line　等密度线
isopycnic surface　等密度面
isostasy　地壳均衡说,均衡代偿学说
isostatic adjustment　均衡调整
isostatic anomaly　均衡异常
isostatic correction　地壳均衡校正
isostatic depression　均衡下降
isostatic disequilibrium　地壳不均衡
isostatic rebound　均衡回弹
isostatic subsidence　均衡下沉
isostatic uplift　地壳均衡抬升

isotac　等解冻线
isotach　等风速线
isotherm　等温线,恒温线
isothermal glacier　等温冰川
isotope　同位素
isotope abundance　同位素丰度
isotope analysis　同位素分析
isotope composition　同位素组成
isotope dating　同位素定年
isotope fractionation　同位素分馏
isotope geochronology　同位素地质年代学
isotope geology　同位素地质学
isotope hydrology　同位素水文学
isotope ratio　同位素比值
isotope standard　同位素标准
isotope stratigraphy　同位素地层学
isotopic discrimination　同位素判别〔值〕
isotopic equilibrium　同位素平衡
isotopic exchange equilibrium　同位素交换平衡
isotopic geochemistry　同位素地球化学
isotopic geothermometer　同位素地热温标
isotopic hydrogeology　同位素水文地质学
isotopic label　同位素标记
isotopic palaeoclimatology　同位素古气候学
isotopic thermometry　同位素测

温法
isotopic tracer 同位素示踪物
isotropic antenna 各向同性天线
isotropic medium 各向同性介质
isotropic reflectivity 各向同性反
射率
isotropic scattering 各向同性散射
isotropic viscous law 各向同性黏
性定律
isotropy 各向同性

J

Jökulhlaup 冰湖突发洪水（冰岛）
Janpanese Earth Resources Satellite 1（JERS-1） 日本地球资源卫星1号
Jet Propulsion Laboratory（JPL） 喷气推进实验室（美国）
Joint Arctic Weather Stations（JAWS） 联合北极天气站
Jordan sunshine recorder 乔唐日照计
Julian calendar 儒略历
Julian day 儒略日
Juneau Icefield （阿拉斯加）朱诺冰原
juvenile water（＝native water） 岩浆水，初生水

K

Köppen-Geiger climate classification 柯本－盖格气候分类
Köppen's climate classification 柯本气候分类
kaavie 卡维雪（苏格兰的大雪）
Kainozoic Era（＝Cenozoic Era） 新生代
Kalman filter 卡尔曼滤波
kame 冰砾阜，冰碛阜
kame delta 冰砾阜三角洲
kame field 冰砾阜群
kame hillock 冰砾阜小丘
kame moraine 冰砾碛
kame plain 冰砾平原
kame plateau 冰砾阜高地,蛇形丘高地
kame ridge 冰砾阜脊,蛇形丘脊
kame terrace 冰砾阜阶地
kame-and-kettle topography 砾阜洼盆地形

Kansas glacial stage　堪萨斯冰期
kar　冰斗
kar cirque　冰斗,冰坑,凹地
karez（＝qanat）　坎儿井,暗渠
karling　冰斗群
karst　喀斯特
karst aquifer　岩溶含水层
karst basin　岩溶盆地
karst glacier　喀斯特冰川（冰川风化侵蚀很像喀斯特地形）
karst plateau　岩溶高原
katabatic cold front　下降〔冷〕锋,下坡锋
katabatic wind　下降风
kavaburd（＝cavaburd）　卡伐布大雪（英国设得兰群岛的大雪）
Kazak Nationality　哈萨克族
Kelvin temperature scale　绝对温标,开尔文温标,K 温标
Kelvin wave　开尔文波
kernel ice　雾凇冰（一种飞机积冰）
kettle hole　锅穴
kettle lake　锅穴湖
kettle moraine　锅穴碛垄
kha-btags（＝hada）　哈达
kiang（*equus kiang*）　藏野驴
killing freeze　（影响植物生长的）冻害
killing frost　严霜

kinamatic positioning　动态定位
kinematic　运动的
kinematic condition　运动学条件
kinematic viscosity coefficient　动黏滞系数
kinematic wave　运动波
kinetic disequilibrium fractionation　动力学非平衡分馏
kinetic energy　运动能
kinetic equation　运动学方程
kinetic isotope effect　动力学同位素效应
kinetic metamorphism　动力变质
Kirgiz Nationality　柯尔克孜族
knife ridge　刃脊
knife-edge crest　刃脊
knob　小丘
knob-and-basin topography　鼓盆地形,羊背石凹凸地形
knot　海里/小时,节
Korean Nationality　朝鲜族
Krakatau　喀拉喀托火山（印尼）
Kriging interpolation　克里金插值法
krill　磷虾
krios　寒冷的（希腊语）
krystic geology　冰雪地质学
Kyoto Protocol（**KP**）　京都议定书

L

La Niña 拉尼娜
La Niña event 拉尼娜事件
labile flow 不稳定流
lacustrine 湖相
lacustrine basin 湖成盆地
lacustrine bog 湖沼
lacustrine civilization 湖居文化（新石器时代）
lacustrine clay 湖积黏土
lacustrine deposit 湖积物,湖泊沉积
lacustrine facies 湖泊相
lacustrine formation 湖泊建造,湖泊层
lacustrine landform 湖泊地形,湖成地形
lacustrine sediment 湖泊沉积,湖相沉积
lake core 湖芯
lake effect snowstorm 大湖效应雪暴
lake ice 湖冰
lake talik 湖泊融区
lake effect snow 湖泊效应降雪
Lambert projection 兰勃特投影
Lambertian reflection 朗伯体反射
lamb-shower （英国）小春雪
laminar flow 层流
laminar motion 层流运动
laminar pattern 层流结构
laminar shear 层流剪切
lamination 层理
Land Cover Classification System（LCCS） 土地覆盖分类系统
land ice 陆地冰
Land Observation Satellite（LOS） 陆地观测卫星
Land Satellite（＝Landsat） 美国陆地卫星
land surface processes 陆面过程
land surface processes parameterization 陆面过程参数化
land surface temperature 陆表温度
land use 土地利用
land use change 土地利用变化
land-derived organic matter 陆源有机质
landfast ice 岸冰,固定冰
landmark 地标,陆标,界桩
land-origin ice 陆源冰
landscape 景观
landscape ecosystem 景观生态系统
landscape map 景观地图
landslide 塌方,山崩,地滑,滑坡
Laplace azimuth 拉普拉斯方位角
Laplace point 拉普拉斯点

lapse line　直减率线
lapse rate　温度垂直梯度，温度直减率
lard ice　脂状冰
Large Aperture Scintillometer (LAS)　大孔径闪烁仪
large floe　大浮冰
large format camera (LFC)　大像幅摄影机
large striated crystals　大条纹状晶体
laser altimeter　激光高度计，激光测高仪
laser distance measuring instrument (=laser ranger)　激光测距仪
laser level　激光水准仪
laser plotter　激光绘图机
laser remote sensing　激光遥感
laser sounder　激光测深仪
laser topographic position finder　激光地形仪
last date　终日
last frost　终霜
last glacial cycle　末次冰期旋迴，末次冰期循环
last glacial maximum (LGM)　末次冰期最盛期，末次冰期冰盛期
last glacial period　末次冰期
last glaciation　末次冰期冰川作用
last ice　终冰
Last Ice Age　末次大冰期，更新世冰期

last ice date　终冰日〔期〕
last interglacial　末次间冰期
Last Interglacial Period　末次间冰期
last snow　终雪
late glacial　晚冰期的，晚（后）冰期
late glacial stage climate　晚冰期气候
late Holocene　晚全新世
late optimum　气候适宜期晚期
late Pleistocene　晚更新世，晚更新世的
late Pliocene　晚上新世
late Quaternary　晚第四纪，晚第四纪的
late Quaternary Glaciation　晚第四纪冰川作用
late spring coldness　倒春寒
late stone age　石器时代晚期
late-glacial period　晚冰期
latent energy　潜能
latent heat exchange　潜热交换
latent heat flux　潜热通量
latent heat transfer　潜热输送
lateral accretion　侧向堆积
lateral corrosion　侧面磨蚀
lateral leaching　侧向淋溶
lateral moraine　侧碛
lateral moraine levee　侧碛堤
lateral moraine loop　侧碛环
latest breakup　最晚解冻日
latest complete freezing　最晚封冻日

latitudinal zonality 纬向地带性
lattice constant 晶格常数
lattice defect 晶格缺陷
Laurentian Shield 劳伦泰地盾
laxation 松弛
layer deformation 层变形
layer echo 层回波
layer reflection 层反射
layer roughness 地层粗糙度
layered cryogenic structure 层状冷生构造
layered cryostructure 层状冷生结构
layered intrusion 层状侵入
layered medium 层状介质
layered permafrost 分层多年冻土
leaching 淋溶
leading beacon 导标
leaf area density 叶面积密度
leaf area index (LAI) 叶面积指数
leeward side (= leeward slope) 背风坡
lens-type cryostructure 透镜状冷生构造
leopard (*panthera pardus*) 金钱豹
level ice 平整冰
level ice thickness 平整冰厚度
level surface 水准面
Lhoba Nationality 珞巴族
lichen 地衣
lichen tundra 地衣冻原
lichenology 地衣学
lichenometry 地衣测年法

light compensation point 光补偿点
Light Detection And Ranging (LiDAR) 激光雷达
light fog 轻雾
light freeze 轻冻
light frost 轻霜
light ice 薄冰(厚0.6 m以下)
light nilas 明尼罗冰,暗冰
light respiration 光呼吸
light saturation point 光饱和点
light snow 小雪
light (ice) floe 薄浮冰
light-frost susceptibility 轻度冻胀敏感性
lignin 木质素
lignosa 木本植被
lily pad ice 饼状冰,莲叶冰
limited area model (LAM) 有限区域模式
limnimeter 水位计
linear reservoir 线性水库
linearly polarized antenna 线性极化天线
liquid chromatography 液相色谱法
liquid limit 液限
liquid phase 液相
liquid water content 液态水含量
Lisu Nationality 傈僳族
lithalsa 石质冻胀丘
lithic anhyorthels 石质脱水正常寒冻土
lithic anhyturbels 石质脱水扰动寒冻土

lithic aquiturbels 石质含水扰动寒冻土
lithic aquorthels 石质含水正常寒冻土
lithic argiorthels 石质黏化正常寒冻土
lithic calci-cryic aridosols 石质钙积寒性干旱土
lithic fibri-permagelic histosols 石质纤维多年冻结有机土
lithic fibristels 石质纤维有机寒冻土
lithic foli-permagelic histosols 石质落叶多年冻结有机土
lithic folistels 石质落叶有机寒冻土
lithic geli-orthic primosols 石质寒冻正常新成土
lithic gypsi-cryic aridosols 石质石膏寒性干旱土
lithic hapli-cryic andosols 石质简育寒性火山灰土
lithic hapli-cryic aridosols 石质简育寒性干旱土
lithic hapli-permagelic gleyosols 石质简育多年冻结纤维土
lithic haplorthels 石质弱育正常寒冻土
lithic haploturbels 石质弱育扰动寒冻土
lithic hemistels 石质半腐有机寒冻土
lithic historthels 石质有机正常寒冻土
lithic histoturbels 石质有机扰动寒冻土
lithic molliturbels 石质暗沃扰动寒冻土
lithic mollorthels 石质暗沃正常寒冻土
lithic psammorthels 石质粗粒正常寒冻土
lithic psammoturbels 石质粗粒正常扰动寒冻土
lithic sapristels 石质高腐有机寒冻土
lithic umbriturbels 石质暗瘠扰动寒冻土
lithic umbrorthels 石质暗瘠正常寒冻土
lithology 岩石学
lithometeor 大气尘粒
lithosphere 岩石圈
lithostratigraphy 岩石层位学,岩石地层学
Little Ice Age（LIA） 小冰期
littoral sediment 滨海沉积物
liverwort 苔类植物
lobate-shaped rock glacier 叶状石冰川
lodgement till 滞碛
loess 黄土
logging 测井
lolly ice 疏松冰,初冰（海冰）
long wave radiation 长波辐射
longitudinal crevasse 纵向裂隙

longitudinal extension 纵向拉伸
longitudinal foliation 纵向叶理
longitudinal profile 纵剖面,纵断面
longitudinal snow patch 纵向雪斑
longitudinal strain rate 纵向应变率
longitudinal wave 纵波
loose avalanche 尘状雪崩,干雪崩
loose snow 疏松雪
low arctic tundra 低北极冻原
low water line 低潮线
lower confining bed 隔水底板
low-frequency radar 低频雷达
low-pass filter 低通滤波
luge 无舵雪橇
luvisolic static cryosol 淋溶静态寒冻土
lysimeter 蒸渗仪

M

macro-scale polygon 巨型多边形
magnetic meridian 磁子午线
magnetic susceptibility 磁化率
major element 主元素
major ions 主要阴阳离子
maladaptation 适应不当
mammoth (= mammuthus) 猛犸象
Manchu Nationality 满族
mandatory level (= mandatory surface) 标准等压面
Manning formula 曼宁公式
Manning roughness 曼宁粗糙率
Manning roughness factor 曼宁粗糙率系数
map compilation 地图编制
map lettering 地图注记
map projection 地图投影
map scale 地图比例尺
mapping accuracy 制图精度
mapping control 图根控制
mapping control point 图根点
marble crust 硬雪壳
marginal channel 边缘河道
marginal crevasse (冰川)边缘裂隙
marginal crushing (冰块碰撞时的)边缘挤轧,浮冰挤轧
marginal ice edge 海冰外缘
marginal ice zone 海冰外缘带
marginal instability 临界不稳定性
marginal lake 冰缘湖
marginal meltwater channel 冰缘融水河道
marginal moraine 边缘冰碛垄,终碛垄,尾碛垄

marginal stream 冰缘河流
marginal terrace 冰缘阶地
marigram 验潮图,验潮记录
marigraph 验潮计
marine hydrological chart 海洋水文图
marine ice layer （冰架底部）海水冻结冰层
marine isotope stage（MIS） 海洋同位素阶段
marine meteorological chart 海洋气象图
Marine Observation Satellite（MOS） 海洋观测卫星
Marine Oxygen Isotope Stage（MOIS） 海洋氧同位素阶段
marine surveying 海洋调查
maritime aerosol 海洋气溶胶
maritime climate 海洋性气候
maritime glacier 海洋性冰川
maritime meteorology 海洋气象学
maritime snow 海洋性积雪
marsh 沼泽,湿地
marsh land 沼泽地
mass balance （冰川的,海冰的）物质平衡,质量平衡（物理）
mass balance year 物质平衡年
mass budget 物质收支
mass concentration 质量浓度（化学）
mass discharge curve 流量累计曲线
mass displacement 块体位移
mass exchange 物质交换
mass flux 物质通量,质量通量

mass loss 物质损失
mass mixing ratio 质量混合比
mass movement （滑坡、泥石流、崩塌等）块体运动
mass spectrometer 质谱仪
mass transfer coefficient 物质输送系数
mass wasting process 物质坡移过程
mass-independent fractionation 非质量同位素分馏
massive agglomerate cryostructure 整体团块状冷生构造
massive cryostructure 整体冷生构造
massive ground ice 大块冰,厚层地下冰
massive movement 块状运动
massive porous cryostructure 整体孔隙冷生构造
master limiting factor 主限制因子
material cycle 物质循环
mattic epipedon 草毡表层
mattic evidence 草毡现象
mattic soil materials 草毡有机土壤物质
maximum carrying capacity 最大承载力
maximum design wave 最大设计波高
maximum diffusion flow 最大扩散流〔动〕
maximum discharge 最大流量

maximum entropy method（MEM）
　　最大熵方法
maximum ice condition　最重冰情
maximum ice thickness　最大冰厚
maximum likelihood estimation
　　（MLE）　最大似然估计
maximum penetration depth　最大
　　穿透深度
maximum possible yield　可能最大
　　产量
maximum sustainable yield（MSY）
　　最大可持续产量
maximum thermometer　最高温
　　度表
meadow　草甸，低湿草地
meadow bog　草甸沼泽
meadow moor　草甸沼泽
meadow peat　草甸泥炭
meadow soil　草甸土
meadow steppe　草甸草原
meadow tundra　草甸苔原
mean annual accumulation　（多）年
　　平均积累
mean annual precipitation　（多）年
　　平均降水量
mean annual temperature　（多）年
　　平均温度
mean anomaly　平均距平
mean deviation　平均偏差
mean error　平均误差
mean gust speed　平均阵风速
mean interdiurnal variability　日际
　　差绝对值的平均值

mean isotherm　平均等温线
mean square error　均方根误差
mean station method　平均站点法
measuring bar　测杆
measuring element　测量要素
measuring mark　测标
mechanical weathering　机械风化
　　〔作用〕
mechanism　机理，机制
medial moraine　中碛，中碛垄
medial moraine bar　中碛堤
medial ridge　中脊
median　中值，中位数
Medieval Climate Optimum（MCO）
　　中世纪气候适宜期
medieval maximum　中世纪极大值
medieval mild phase　中世纪温暖期
Medieval Warm Epoch（MWE）
　　中世纪暖期
Medieval Warm Period（MWP）
　　中世纪暖期
medium lake ice　中等湖冰（厚度
　　15～30 cm）
Medium Resolution Imaging Spectrometer（MERIS）　中分辨率
　　〔成像〕光谱仪
medium winter ice　中等冬冰，薄
　　冬冰
medium-frost susceptibility　中度
　　冻胀敏感性
megadune　大雪丘，沙山
megainterstadial　大间冰阶
Megathermal Period　（全新世）大

暖期
Meighen Ice Cap （加拿大）米恩冰帽
meiotherm 低温植物
meltout moraine 融出冰碛垄
meltout till 融出碛
melt period 解冻期,消融期
melt pond （海冰面上的）融水池
melt season 融化季节
melt water 融水
melt-freeze metamorphism 冻融变质作用
melting 融化
melting band 融化带
melting date 融冰日
melting ice 融冰
melting layer 融化层
melting level 融化层,融化高度
melting point 融化点
meltwater runoff 融水径流
membrane filter 滤膜
membrane permeability 膜透性
membrane stress 薄膜应力
Mercator projection 墨卡托投影
mercury barometer 水银气压表
mercury thermometer 水银温度表
mercury-in-glass psychrometer 玻璃管水银干湿表
mercury-in-glass thermometer 玻璃管水银温度表
mercury-in-steel thermograph 钢管水银温度计
meridian 子午线
meridian circle 子午圈,子午环
meridian plane 子午面
meridional circulation 经向环流
meridional flow 经向气流,经向海流
meridional gradient 经向梯度
meridional overturning circulation （MOC） 经向反转环流
mesic organic cryosol 中湿有机寒冻土
mesochthonophyta 内陆植物
mesochthonophytia 内陆植物群系
Meso-Proterozoic Great Ice Age （MPGIA） 中元古代大冰期
mesoscale circulation 中尺度环流
mesoscale motion 中尺度运动
mesotherm plant 温带植物
mesothermal climate （＝temperate climate） 温带气候
mesothermophytia 温带植物群落
metabolic rate 代谢速率
metabolism 新陈代谢
metamorphism 变质作用
metamorphose 变质
meteorite（＝meteorolite） 陨石
meteorological balloon 探空气球
meteorological disaster 气象灾害
meteorological element 气象要素
meteorological instrument 气象仪器
meteorological observation 气象观测
meteorological observatory 观象台
meteorological phenomenon 气象现象,天气现象

meteorological record 气象记录
meteorological satellite 气象卫星
Meteorological Satellite (METEOSAT)
　　欧洲同步气象卫星
meteorological station 气象站
meteorological symbol 气象符号
meteorological telecommunication hub 气象通信枢纽
meteorological tide 气象潮
meteorological visibility 气象能见度
meteorologist 气象工作者,气象学家
meteorology 气象学
methane bacteria 甲烷细菌
methane clathrate (=natural gas hydrate) 天然气水合物,可燃冰
methane source 甲烷源
methane (CH$_4$) 甲烷
methanesulfonic acid (MSA) 甲基磺酸
methanogenesis 甲烷生成作用
methanogens 甲烷生成菌
methyl 甲基
methylsulfonic acid (MSA) 甲烷磺酸
microbe 细菌,微生物
microbial action 微生物作用
microbial activity 微生物活动
microbial biomass 微生物量
microbial decomposition 微生物分解
microclimate 小气候
microcopying 微缩拷贝
microenvironment (=microhabitat) 小生境,小环境
microfilm map 缩微地图
microfilming 缩微摄影
microflora (=microbiota) 微生物区系
microgravimetry 微重力测量
microhabitat (=microenvironment) 小生境,小环境
microorganism (=microbe) 微生物
microparticle 微粒
microphotography 缩微摄影
microrheology 微流变学
micro-scale polygon 小型多边形
microscopic analysis 显微分析
microscopic crack 显微裂缝
micro-section 显微切片
microsequence 微层序
microvarve 极细纹泥
micro-void 微空穴(冰晶)
microwave 微波
microwave altimeter 微波高度计
microwave hologram radar 微波全息遥感
Microwave Humidity Sounder (MWHS) (风云三号卫星的)微波湿度辐射计
microwave image 微波图像
microwave imagery 微波图像
microwave radar 微波雷达
microwave radiation 微波辐射
microwave radiometer 微波辐射计

Microwave Radiometer Imager（MWRI）（风云三号卫星的）微波成像仪
microwave remote sensing 微波遥感
microwave scatterometry 微波散射测量
microwave sensor 微波传感器
microwave sounding unit（MSU）微波探测器
microwave spectrometer 微波频谱仪
microwave spectrum 微波谱
microwave-infrared sounding 微波—红外探测
middle infrared 中红外
middle latitude 中纬度
mid-latitude westerlies 中纬度西风带
Mid-Pleistocene Transition（MPT）中更新世气候转型
Mie scattering 漫散射
migration 迁移,迁徙
Milankovitch cycle 米兰科维奇循环,米兰科维奇周期
Milankovitch hypothesis 米兰科维奇假说
Milankovitch oscillation 米兰科维奇振荡
Milankovitch theory 米兰科维奇理论,米氏理论
mild clay 亚黏土
milk ice（=milky ice） 乳冰,乳色冰
millennial scale 千年尺度
millennial variability 千年变率
millennial-scale variation 千年尺度变化
millibar barometer 毫巴(标尺)气压表
millibar（mb） 毫巴(气压单位,1 mb=1 hPa)
millimetre of mercury 毫米汞柱(1毫米汞柱=1.333 hPa)
Mindel Glaciation （欧洲阿尔卑斯）民德冰期
Mindel-Riss interglacial stage 民德—里斯间冰期
mineral dust aerosol 沙尘气溶胶
mineralization 矿化度,矿化作用
mineralization rate 矿化速度
minerogenic palsa 矿质泥炭丘
minimum quadrat area 最小样方面积
minimum quadrat number 最小样方数目
minimum thermometer 最低温度表
miniski 小雪橇
minor cycle 小循环
minor elements 微量元素
minority nationality 少数民族
Miocene Epoch 中新世
Miocene Series 中新统
miothermic period 温和期(指间冰期)(旧称)
mirage 海市蜃楼
missing carbon sink 失踪碳汇
missing data 缺测数据
mist 雾
mitigation 减缓,减排

mixed climbing （冰岩）混合攀登
mixed cloud （冰水）混合云
mixed layer 混合层
mixed pixel 混合像元
mixed precipitation 雨夹雪
mixing layer 混合层
mixing ratio 混合比
model calibration 模型率定，模型标定
model error 模型误差
model experiment 模拟试验
model parameter 模型参数
model recalibration 模式修正
model resolution 模拟分辨率
model sensitivity 模式敏感性（度）
model structure 模型结构
model validation 模式验证
model verification 模型检验
modeling 模拟，建模
moderate fog 中雾
moderate rain 中雨
Moderate Resolution Imaging Spectroradiometer （MODIS） 中分辨率成像光谱仪
moderate sea 中浪（风浪 3 级）
modulation frequency 调制频率
modulation transfer function （MTF） 调制传递函数
modulation waveform 调制波形
modulator 调制器
module 模块
Mohr stress circles 莫尔应力圆
Mohr-Coulomb failure criteria 莫尔—库仑破坏准则
moire 龟纹
moist convection 湿对流
moist snow 湿雪，潮雪
moist snow-flake 湿雪花
moisture 水分,水汽
moisture budget 水汽收支
moisture content 含水量
moisture equivalent 持水当量（土壤）
moisture migration 水分迁移
moisture retention curve 持水曲线
molar gas constant 摩尔气体常数
mole 摩尔
molecular 分子的
molecular diffusion 分子扩散
molecular diffusion coefficient 分子扩散系数
molecular scattering 分子散射
mollic epipedon 暗沃表层
mollic hapli-permagelic gleyosols 暗沃简育多年冻结纤维土
mollition 解冻，活动层融化过程（土壤学）
molliturbels 暗沃扰动寒冻土
mollorthels 暗沃正常寒冻土
momentum 动量
momentum conservation 动量守恒
momentum flux 动量通量，涡动动量通量
momentum moisture flux 涡动水汽通量
momentum transfer 涡动动量输

送，动量传输
Monba Nationality 门巴族
Mongol Nationality 蒙古族
Mongolian high 蒙古高压
monimolimnion （湖水）永滞层
monitor 监测
monitoring network 监测网
monochromatic albedo 单频反照率
monochromatic radiation 单色辐射
monochromatic reflectance 单色反射率
monoporate （孢粉）单孔的
monopulse radar 单脉冲雷达
monostatic radar 单基雷达
monsoon 季风
monsoon circulation 季风环流
monsoon current 季风海流
monsoon season 季风季
monthly mean temperature 月平均温度
morainal apron（＝morainic apron） 冰碛裙
morainal delta 冰碛三角洲
morainal lake 冰碛湖
moraine 冰碛垄
moraine belt 冰碛带
moraine breccia 冰碛角砾岩
moraine dam（＝morainic dam） 冰碛坝
moraine fan 冰碛扇
moraine height-to-width ratio 冰碛坝高宽比
moraine hill 冰碛丘陵
moraine inside flank 冰碛坝迎水坡
moraine kame 冰碛阜
moraine lake 冰碛湖
moraine land 冰碛地
moraine loop 冰碛环
moraine plain 冰碛平原
moraine plateau 冰碛高原
moraine soil 冰碛土
moraine terrace 冰碛阶地，冰碛台地
moraine-covered glacier 冰碛覆盖型冰川，土耳其斯坦型冰川
moraine-dammed lake 冰碛阻塞湖
morainic debris 冰碛岩屑
morainic deposit 冰碛沉积，冰碛层
morainic fan 冰碛扇
morainic loop 冰碛环
morainic material 冰碛物质
morainic topography 冰碛地形，冰碛地貌
morass 沼泽
morphogenesis 地貌成因〔学〕，地形成因〔学〕，地貌发生
morphographic landscape 地貌景观
morphologic(al) 形态〔学〕的，地貌的
morphostratigraphy 地貌地层学
mosaic community 镶嵌群落
mosaic distribution 镶嵌分布
mosaic structure 镶嵌结构

mosaic succession 镶嵌演替
mosaic vegetation 镶嵌植被
mositure capacity 持水量
mositure meter （土壤）湿度计
moss 苔藓,地衣
moss original tape 青苔苔原带
Mosuo Nationality 摩梭族
mother plant 母株
mottlic geli-alluvic primosols 斑纹寒冻冲积新成土
mottlic halpi-gelic cambosols 斑纹简育寒冻雏形土
mottlic molli-gelic cambosols 斑纹暗沃寒冻雏形土
mottlic umbri-gelic cambosols 斑纹暗瘠寒冻雏形土
mouldability 可塑性
moulin kame 冰瓯阜
moulin（＝glacier well） 冰川竖井
mountain barrier 山岳屏障,山地障碍
mountain bike（MTB） 山地自行车
mountain birch（betula tortuosa） 弯桦
mountain breeze 山风
mountain cap cloud 山顶云
mountain climate 山地气候
mountain climatology 山地气候学
mountain ecotourism 山地生态旅游
mountain fog 山雾
mountain forcing 地形强迫作用,山脉强迫作用
mountain forms 山形
mountain glaciation 山地冰川作用
mountain glacier 山地冰川
mountain meadow soil 山地草甸土
mountain permafrost 山地多年冻土
mountain physiologic effect 高山生理反应
mountain range 山脉
mountain rescue 高山救援
mountain ridge 山脊
mountain sick（＝altitude sick） 高山病
mountain ski tourism 山地滑雪旅游
mountain sport 山地体育运动
mountain torrent 山洪
mountain tourism 山地旅游
mountain tundra 高山苔原
mountain valley 山谷
mountain weather 山地天气
mountain wind 山风
mountaineer 登山运动员
mountaineering 登山
mountain-gap wind 山口风
mountain-making movement（＝orogeny） 造山运动
mountain-plain circulation 高山平原间环流
mountain-plain wind systems 山地平原风系
mountain-valley wind（＝mountain-valley breeze） 山谷风
mountain-valley wind systems 山谷风系

mountain-wave cloud 山地波状云
moutonnee 冰川擦痕,羊背石
moving boundary condition 移动边界条件
moving ice 移动冰,活动冰
mud boil 翻浆,泥沸
mud circle 泥环
mud crack 泥裂
mud flow deposit 泥流沉积
mud flow(＝mudflow) 泥流
Mud Glacier 玛德冰川(加拿大)
mud hummock 泥丘
mud line 泥线
mud polygon 泥质多边形
mud spring 泥泉
mud volcano 泥火山
muddy sand 泥质砂
muddy sorted circle 泥质分选环
muddy volcanic cone 泥火山锥
mudflow terrace 泥流阶地
mudstone 泥岩
Multi-angle Imaging Spectro Radiometer(MISR) 多角度成像光谱仪
multi-annual storage 多年库容,多年蓄水量
multi-band remote sensing 多波段遥感
multi-baseline inteferometry 多基线干涉
multi-beam sounding system 多波束测深系统
Multicolor Spin-Scan Cloud Cover Camera(MSSCC) 多色自旋扫描云覆盖摄影机
Multidisciplinary Earth Observation Satellite(MEOS) 多用地球观察卫星
Multifrequency SAR(MSAR) 多频率合成孔径雷达
Multi-functional Transport Satellite(MTSAT) 日本的新一代静止气象卫星
Multilateral Recognition Arrangement(MLA) 多边协议
multi-look image 多视图像
multi-phase flow 多相流
multiple correlation 复相关
multiple equilibria 多平衡态
multiple scattering 多次散射
multi-point method 多点测速法
multi-polarization 多极化
multi-spectral 多光谱
multi-spectral classification 多光谱分类
multi-spectral image 多谱段影像
multi-spectral imager 多光谱成像仪
Multi-Spectral Imager(MSI) (欧空局 Sentinel-2 卫星的)多光谱成像仪
multi-spectral scanner(MSS) 多波段扫描仪
multi-temporal 多时相
multi-temporal analysis 多时相分析
multi-temporal remote sensing 多

时相遥感
multi-year-ice 多年冰（海冰）
multi-year ice floe 多年浮冰块（海冰）
multi-year landfast sea ice 多年陆缘海冰
musk deer（*moschus moschiferas*）

原麝
muskeg （有机质多,尤指北美北部和北欧的）泥炭沼泽,厚苔沼泽,泥炭藓沼泽
Muskingum method 马斯京根法
muskoxen (*ovibos moschatus*) 麝牛

N

nabivnoy ice 多层冰,筏状冰
naled(pl. naledi) 冰椎(俄语)
nanospore （孢粉）微孢子
Nansen Ice Shelf 南森冰架(南极)
narrow passage 隘道
NASA Scatterometer instrument (NSCAT) 美国航空航天署散射计
National Inventory Report (NIR) 国家清单报告
national nature reserve 国家自然保护区
National Oceanic Satellite System (NOSS) 国家海洋卫星系统
national park 国家公园
National Polar-orbiting Operational Environmental Satellite System (NPOESS) 美国环境监测业务极轨卫星系统
nationality composition 民族结构,民族构成

natric argiorthels 碱化黏化正常寒冻土
natural assets 自然资产
natural background 自然本底,自然基准
natural boundary condition 自然边界条件
natural capital 自然资本
natural coordinates 自然坐标
natural ecosystem 自然生态系统
natural fertility 自然肥力
natural flow 天然水流
natural forcing 自然驱动力
natural forest regeneration 森林自然更新
natural gas hydrate (=methane clathrate) 天然气水合物,可燃冰
natural gas pipeline 天然气管道
natural groundwater regime 天然地下水状态
natural ice 天然冰

natural landscape	自然景观
natural radioelement	天然放射性元素
natural resources	自然资源
natural scenery	自然风景
natural selection	自然选择
natural succession	自然演替
natural sulfur cycle	自然硫循环
natural synoptic region	自然天气区
natural synoptic season	自然天气季节
natural turnover rate	自然周转率
natural vegetation	自然植被
natural wavelength	固有波长
nature conservation	自然保护
nature reserve	自然保护区
Navy Navigation Satellite System (NNSS)	海军导航卫星系统
Navy Ocean Surveillance Satellite (NOSS)	海军海洋监视卫星
Naxi ancient music	纳西古乐
Naxi Nationality	纳西族
Neanderthal man	尼安德特人
near infrared	近红外
nearest neighbor	最邻近法
nearshore current system	近岸流系
near-shore ice	（河、湖、海冰）近岸冰
nearshore zone	近滨带,近海,近岸带
near-surface heat-flow	近表面热流
Nebraskan Glacial Stage	内布拉斯加冰阶段
nebula	星云
needle ice (＝pipkrake)	针状冰,冰针
needle ice creep	针状冰蠕变
needle-leaved evergreen forest	针叶常绿林
negative balance	负均衡,负平衡
negative feedback	负反馈
negative ion (＝anion)	负离子,阴离子
negative mass balance	负物质平衡
neighborhood committee	居民委员会
Neogene Period	新第三纪
Neogene System	新第三系,新近纪
neoglaciation	新冰期冰川作用
Neo-Ice Age	新冰期
neolith	新石器
neolithic	新石器时代的
neolithic age	新石器时代
neolithic culture	新石器文化
neolithic period	新石器时代
neoorogenic zone	新造山带
Neo-Proterozoic Great Ice Age (NPGIA)	新元古代大冰期
Neozoic Era	新生代
nesting problem	（嵌)套网格问题
net ablation	净消融〔量〕,纯消融〔量〕
net accumulation	净积累〔量〕,纯积累〔量〕

net advection 净平流,纯平流
net balance 净平衡,纯平衡
net balance isoline 净平衡等值线,纯平衡等值线
net biome production（NBP） 净生物群区生产量
net biome productivity（NBP） 净生物群区生产力
net ecosystem exchange（NEE） 净生态系统交换
net ecosystem production（NEP） 净生态系统生产量
net ecosystem productivity（NEP） 净生态系统生产力
net flow 净流〔量〕
net impact 净影响
net photosynthesis 净光合作用
net photosynthetic rate 净光合作用速率
net primary productivity（NPP） 净初级生产力
net pyrgeometer 大气净辐射表,净天空辐射表
net pyrradiometer 辐射平衡表
net radiation 净辐射
net radiation flux 净辐射通量
net radiometer 净辐射表
net solar radiation 净太阳辐射
net terrestrial radiation 净地面辐射
network of watersourses 水网,水道网
neural network 神经网络

neutron capture 中子捕获
neutron flux 中子流量
neutron moisture gauge 中子含水量测定器
neutron moisture meter 中子（土壤）含水量仪
névé（＝firn） 粒雪（法）
new ice （河、湖、海）新冰
new snow 新雪,新降雪,新雪（雪晶形状可辨认或 24 h 内新降的雪）
new snow-ice 新雪冰（落在水面的雪所形成）
New Stone Age 新石器时代
Newton Viscosity Law 牛顿黏性定律
Newton's Law of Cooling 牛顿冷却定律
Newtonian cooling 牛顿冷却
Newtonian fluid 牛顿流体
Newtonian shear viscosity 牛顿剪切黏度
Newtonian viscous fluid 牛顿黏性流体
nexine （孢粉）外壁内存
niche 雪凹
niche differentiation 生态位分化
niche diversification 生态位多样化
niche glacier （山坡）凹地冰川
niche intersection 生态位交叉
niche overlap 生态位重叠
niche partitioning 生态位分割

niche relationship　生态位关系
niche size　生态位大小
niche space　生态位空间
niche structure　生态位结构
Nickerson Ice Shelf　尼克尔森冰架（南极）
nieve penitente　雪塔林（南美），冰塔林
Nihewan fauna　泥河湾动物群
Nihewan formation　泥河湾组
Nihewanian stage　泥河湾阶
nila(s)（＝nillah）　尼罗冰（平滑水平的薄冰，厚度小于 10 cm），冰壳,冰皮,冰衣
Nimbus　（美国）气象卫星
Nimbus-E Microwave Spectrometer（NEMS）　雨云—5 号微波频谱仪
nip　冰挤压,夹住航船等
nipher shield　奈弗防护罩（雨量器用）
nitrate　硝酸盐
nitrate ion　硝酸根离子
nitrate nitrogen　硝态氮
nitrate radical　硝酸根
nitrate reduction　硝酸盐还原作用
nitration　硝化
nitric acid　硝酸
Nitric Acid Trihydrate（NAT）　硝酸三水合物
nitric anhyorthels　亚硝酸脱水正常寒冻土
nitric anhyturbels　硝酸盐脱水扰动寒冻土
nitric oxide（NO）　一氧化氮
nitrification　硝化作用
nitrite　亚硝酸盐
nitrite reductase　亚硝酸盐还原酶
nitrogen assimilation　氮同化〔作用〕
nitrogen balance　氮平衡
nitrogen cycle　氮循环
nitrogen deficiency　缺氮
nitrogen deposition　氮沉降
nitrogen dioxide（NO_2）　二氧化氮
nitrogen fixation　固氮作用,固氮
nitrogen fixing bacteria　固氮菌
nitrogen oxides（NO_x）　氮氧化物（N_2O,NO,NO_2 等）
nitrogen pentoxide（N_2O_5）　五氧化二氮
nitrogen use efficiency　氮利用效率
nitrogen（N）　氮（气）
nitrogenase　固氮酶
nitrogen-containing amines　含氮氨
nitrous acid　亚硝酸
nitrous oxide（N_2O）　氧化亚氮,一氧化二氮
nival　雪的（拉丁）
nival belt　雪带
nival climate　冰雪气候
nival erosion　雪蚀
nival flood（＝snowmelt flood）　融雪洪水,春汛
nival zone　雪带
nivation（＝snow erosion）　雪蚀,

霜蚀
nivation bench 雪蚀条状台地
nivation cirque 雪蚀冰斗
nivation depression 雪蚀洼地
nivation glacier 雪蚀冰川
nivation hollow 雪蚀洼地
nivation terrace 雪蚀阶地
nivation trough valley 雪蚀槽谷
niveo-aeolian deposit 雪蚀,风成堆积
niveo-erolian deposit 雪风沉积
nivological phenomena 积雪现象
nivometer 超声雪深测量仪
nivometric coefficient 雪雨比
non-methane hydro-carbons（NMHCS）非甲烷烃,非甲烷碳氢化物
no regrets choice 无悔选择
no regrets policy 无悔政策
nocturnal boundary layer（NBL）夜间边界层
no-frost zone 无霜区
noise equivalent temperature difference（NETD）噪声等效温差
nomad 游牧民
nomadic culture 游牧文化
nondimensionalization 无量纲化
nonequilibrium state 非平衡态
nonfrost period 无霜期
Non-Governmental Organization（NGO）非政府组织
nonindigenous 非本地的
nonlinear regression 非线性回归
non-methane hydrocarbons（NMHCs）非甲烷烃
nonrenewable resource 不可再生资源
non-sea-salt 非海盐
non-sorted circle 非分选环
non-sorted polygon 非分选多边形
nonsorted stripe 非分选条
normal distribution 正态分布
normal fault 正断层
normal incidence 正入射,垂直入射
normal section 法截面
normal strain 正应变
normal strain rate 正应变率
normal stress 正应力
normal stress component 正应力分量
normalization 归一化,正态化,归一法,标准化
normalized difference snow and ice index（NDSII） 归一化雪冰指数
normalized difference snow index（NDSI） 归一化积雪指数
normalized difference vegetation index（NDVI） 归一化植被指数
normalized difference water index（NDWI） 归一化水体指数
normalized difference cloud index（NDCI） 归一化云指数
North Atlantic current 北大西洋洋流
North Atlantic drift 北大西洋漂流

north atlantic fin whales (*balaenoptera physalus*) 鳍鲸
North Atlantic Oscillation (NAO) 北大西洋涛动
North Atlantic stream 北大西洋海流
north frigid zone 北寒带,北极带
North Greenland Eemian Ice Core Drilling Project (NEEM) 格陵兰北部埃姆冰芯计划
north magnetic pole 磁北极
North Pacific Oscillation (NPO) 北太平洋涛动
north polar region 北极地区
North Pole 北极点
north temperate zone 北温带
Northern Hemisphere 北半球
Northern Hemisphere Glaciation 北半球冰川作用
northern latitude 北纬
northern lights (=aurora) 北极光
northern pike (*esox reicherti*) 北方狗鱼
northern small limestone moss (*seligeria polaris*) 极地细叶藓
northern vernal equinox 春分,春分点
northern winter 北半球冬季
northerners 北方人

no-till farming 免耕农业
Novaya Zemlya 新地岛
nuclear dust 核尘
nuclear explosion 核爆炸
nuclear test 核试验
nuclear winter 核冬天
nucleation 成核作用
nuclei acids 核酸
nuclei bubble 核泡
nuclepore filter 核孔滤膜
nudation 裸露
number concentration 数浓度
numerical calculation 数值计算
numerical computation 数值计算
numerical experiment 数值试验
numerical forecast 数值预报
numerical model 数值模式
numerical scheme 数值方案
numerical simulation 数值模拟
numerical solution 数值解
nunatak 冰原岛峰,冰原岛山
nutation 章动
nutation of obliquity 交角章动
nutation period 章动周期
nutrient fluid (=nutrient solution) 培养液
nutrient solution (=nutrient fluid) 培养液
Nyenyexonglha Glaciation 聂聂雄拉冰期

O

oasis 绿洲
oasis development 绿洲开发
oasis effect 绿洲效应
oasis farming 绿洲农业
oasis soil 绿洲土壤
oblique bedding 斜层理
obliquity 倾斜度
observation target 测量觇标
observation technology 观测技术
observation well 验潮井,观测井
observational network 测站网
observatory 观象台,天文台
ocean basin 大洋盆地,洋盆
ocean circulation 大洋环流,洋流
ocean climate 海洋气候
ocean current 海流,洋流
ocean current system 洋流系统
Ocean Drilling Program（ODP） 海洋钻探计划
ocean floor 洋底
Ocean Satellite（＝Oceansat） 印度海洋卫星
ocean sedimentation rate 海洋沉积速率
ocean sounding chart 大洋水深图
ocean surface current 表层海流
ocean-atmosphere heat exchange 海气热交换
ocean-atmosphere interaction 海气相互作用,海气作用
ocean-atmosphere interchange 海气交换
oceanic gyre 大洋涡旋
oceanic tide 大洋潮汐
oceanic turbulence 大洋湍流,大洋涡动
oceanic upwelling 大洋上升流
oceanography 海洋学
off-sea wind（＝on-shore wind） 向岸风
offset 偏移,位移
offset policy 补偿政策
off-shore current 离岸流
offshore permafrost 海（湖）底多年冻土
off-shore wind 离岸风
ogive 冰川弧拱
old bay ice 老（海或湖的）湾冰
old ice （河、湖、海）老冰,多年冰
old snow 老雪
older Dryas 中仙女木期
oldest Dryas 老仙女木期
old-field ecosystem 弃耕地生态系统
olefine 烯烃
Oligocene 渐新世
oligomerization reaction 低聚反应
oligotropic plant 贫养植物

ombrograph 微雨量计
ombrometer 微雨量器
Omega chart 奥米伽海图
omnidirectional antenna 全向天线
omnidirectional radiometer 全向辐射仪
one-way grid nesting 单向〔嵌〕套网格
one-year ice 一年冰,当年冰
onshore current 向岸流
onshore permafrost 岸区多年冻土
opaque ice 不透明冰
open boundary 开放边界
open channel flow 明渠水流
open community 开放群落
open drift ice (= open pack ice) 稀疏流冰群(冰量 4/10～6/10 或 3/8～5/8)
open harbour 不冻港
open ice 稀疏冰
open ice-edge 稀疏浮冰边缘(线)
open inlet 不冻河口,开敞河口
open pack ice (= open drift ice) 稀疏浮冰群(冰量 4/10～6/10 或 3/8～5/8)
open sea 不冻海;公海
open system pingo 开放型冰丘,开放型冻胀丘
open talik 贯穿融区
open water 无冰水面
open-cavity ice 开敞洞穴冰
open-path eddy covariance (OPEC) 开路涡度相关

open-system freezing 开敞系统冻结
opportunity cost (经)机会成本
optical axis 光轴
optical graphical rectification 光学图解纠正
optical hygrometer 光学湿度表
optical image processing 光学图像处理
optical instrument positioning 光学(仪器)定位
optical level 光学水准仪
optical path 光程,光学路径
optical pattern recognition 光学图像识别
optical properties 光学特性
optical rectification 光学纠正
optical refractive index 光学折射率
optical remote sensing 光学遥感
optical scattering 光散射
optical sensor 光学传感器
optical snow depthmeter 雪深光学测量仪
optical theodolite 光学经纬仪
optical thickness 光学厚度
optical transfer function (OTF) 光学传递函数
optically stimulated luminescence dating (OSL dating) 光释光测年
optimum embankment height 最佳路堤高度
optimum insulation thickness 最佳保温层厚度
optimum population 适度人口

optoacoustic detection 光声探测
orbiculic cryogenic fabric 冷生环状微结构
orbital altitude 轨道高度
orbital eccentricity 轨道偏心率
orbital eccentricity variation 轨道偏心率变化
orbital forcing 轨道驱动
orbital inclination 轨道倾角
orbital parameter 轨道参数
orbital variation 轨道变化
orbiting geophysical observatory (OGO) 轨道地球物理观测
ordinary climatological station 一般气候站
Ordovician-Silurian Great Ice Age 奥陶纪—志留纪大冰期
ore 矿,矿石
organic 有机的,生物的
organic acids 有机酸
organic aerosol 有机气溶胶
organic agriculture (= organic farming) 有机农业
organic carbon 有机碳
organic cryosol 有机寒冻土
organic epipedons 有机表层(类)
organic geochemistry 有机地球化学
organic matter 有机质
organic nitrate 有机硝酸盐
organic peroxides 有机过氧化物(通式 ROOR 或 ROOH)
organic pollutant 有机污染物
organic soil materials 有机土壤物质
organic-mud zone 有机质泥带
organism 有机体,生物
organogenic 有机成因的
orientated section 定向切片
orientation 取向,定向,走向
orientation connection survey 定向连接测量
orientation point 定向点
oriented overgrowth (= epitaxis) (晶体)定向生长,外延生长
oriented thaw lake 定向热融湖
original-fissure water 原生裂隙水
orogeny (= mountain-making movement) 造山运动
orographic 地形的,山地的
orographic downward wind 地形下坡风
orographic drag 地形拖曳
orographic effect 地形效应
orographic forcing 地形作用力,地形强迫
orographic precipitation 地形降水
orographic rainfall 地形雨
orographic snowline 地形雪线
orographic wave 地形波
Oroqen Nationality 鄂伦春族
orthels 正常寒冻土亚纲
orthic dystric static xryosol 正常不饱和静态寒冻土
orthic dystricturbic cryosol 正常不饱和扰动寒冻土
orthic eutric static xryosol 正常饱

和静态寒冻土
orthic eutric turbic cryosol 正常饱和扰动寒土
orthogonal transformation 正交变换
orthogonalization 正交化
orthographic projection 正射投影
orthophoto stereomate 正射影像立体配对片
orthorectification 正射校正
orthoscope 正射投影仪
oscillation 振动,振荡,涛动
ott (current) meter 奥特流速仪
oued 枯水河
outburst flood 突发洪水,溃决洪水
outer boundary layer (OBL) 外边界层
outgoing radiation 向外辐射
outlet glacier 溢出冰川
outwash 冰水沉积
outwash apron 冰水沉积扇
outwash area 冰水沉积区
outwash coast 冰水沉积海岸
outwash delta 冰水沉积三角洲
outwash deposit 冰水沉积物
outwash depositional plain 冰水沉积平原
outwash fan 冰水扇
outwash gravel 冰水砾
outwash plain 冰水平原,外冲平原
outwash stream 冰水河流
outwash terrace 冰水阶地

oval-shaped belt 极光卵带
overburden pressure 覆盖层自重压力,超载压力
overdeepen 过量下蚀
overflow 溢流
overgrazing 过度放牧
overland flow 地表漫流
overlying layer 上覆层
oversaturation 过饱和
overthrusting 掩冲
overturned fold 倒转褶皱
overwinter 越冬
oxbow lake deposit 牛轭湖沉积
oxidant 氧化剂
oxidation 氧化作用
oxidation reaction 氧化反应
oxidation state 氧化态
oxidation-reduction 氧化还原反应
oxide 氧化物
oxidizing environment 氧化环境
oxygen 氧
oxygen content 氧气含量
oxygen cylinder 氧气瓶
oxygen isotope 氧同位素
oxygen isotope ratio 氧同位素比率
oxygen isotope record 氧同位素记录
oxygen partial pressure 氧气分压
oxygen saturation 氧饱和
oxygen stable isotope 稳定氧同位素
ozone 臭氧

ozone concentration 臭氧浓度
ozone depleting substance 臭氧耗减物质
ozone depletion 臭氧损耗
ozone flux 臭氧通量
ozone hole 臭氧洞
ozone isopleth 臭氧等值线
ozone layer 臭氧层
ozone photochemistry 臭氧光化学
ozone reaction 臭氧反应
ozone shield 臭氧防护层
ozone-depleting potential（ODP）臭氧耗减势,臭氧层破坏潜力
ozonesonde 臭氧探空仪
ozone-temperature sensor 臭氧－温度传感器
ozonide 臭氧化物
ozonization 臭氧化过程
ozonograph 臭氧测量仪
ozonometer 臭氧计
ozonometry 臭氧测定术
ozonopause 臭氧层顶
ozonoscope 臭氧测量仪
ozonosphere 臭氧层

P

Pacific anticyclone 太平洋反气旋
Pacific decadal oscillation（PDO）太平洋年代际振荡
pack ice 浮冰群
pack ice zone 浮冰区
packed slush 雪浆
pagoscope 测霜仪
pal(a)eotemperature 古温度
pal(a)eotemperature record 古温度记录
palaeoanthropology 古人类学
palaeobiotope 古生境
palaeochannel 古河床,古河道,古河槽
palaeoclimate 古气候
palaeoclimate reconstruction 古气候重建
palaeoclimatology 古气候学
palaeo-equilibrium line altitude 古平衡线高度
Palaeogene 早第三纪,早第三纪的
Palaeogene period 早第三纪
Palaeogene system 下第三系
palaeoglaciation 古冰川〔作用〕
palaeoglciation record 古冰川记录
palaeo-ice stream 古冰流
palaeo-ice thickness 古冰厚
palaeolimnology 古湖沼学
palaeolith 旧石器
palaeolithic 旧石器时代的
palaeomagnetic dating 古地磁测年
palaeomagnetic field 古地磁场

palaeomagnetic pole 古磁极
palaeomagnetic time scale 古地磁年代表
palaeomagnetism 古地磁学,古地磁
palaeomonsoon 古季风
palaeontologist 古生物学家
palaeo-ocean 古大洋,古海洋
palaeopalynology 古孢粉学
palaeopedological 古土壤的
palaeo-periglacial action 古冰缘作用
palaeophytogeography 古植物地理学
palaeophytologist 古植物学家
palaeo-snowline 古雪线
palaeosol 古土壤
palaeotemperature curve 古温度曲线
palaeotemperature variation 古温度变动
palaeothermometry 古温度测定法
Palaeozoic Era 古生代
Paleo-Equilibrium Line Altitude (Paleo-ELA) 古平衡线高度
Paleo-Proterozoic Great Ice Age (PPGIA) 古元古代大冰期
palevent 古地理事件
palobiogeography 古生物地理学
palsa 泥炭丘
palynoevent 孢粉演变
palynofacies 孢粉相
palynoflora 孢粉植物群
palynogram 孢粉图式,孢粉图解
palynology 孢粉学
palynonorph 孢粉体,孢粉型
pancake ice 饼冰
Panchromatic Remote-sensing Instrument for Stereo Mapping (PRISM) 立体制图全色遥感器
pan-ice 饼冰(海冰)
panoramic radar 宽式雷达
paracalicic hapli-cryic aridosols 弱钙简育寒性干旱土
paraglacial deposit 冰缘沉积
paraglacial fan 冰缘扇
paraglacial geomorphology 冰缘地貌
paraglacial landscape 冰缘景观
paraglacial sediment 冰缘沉积
paraglacial terrace 冰缘阶地
paragypsic hapli-cryic aridosols 弱石膏简育寒性干旱土
parameterization 参数化
parameterized model 参数化模式
parametric model 参数模型
partial pressure 分压
participatory rural appraisal (PRA) 参与式农村评估
particle size distribution 粒度分布
particulate organic carbon 颗粒有机碳
parts per billion (ppb) 十亿分率
parts per million (ppm) 百万分率
parts per trillion (ppt) 万亿分率

passive absorption 被动吸收
passive microwave 被动微波
Past Global Changes（PAGES） 过去全球变化计划（IGBP 的核心计划之一）
pastoral region 牧区
pasture 牧场
patchy snow cover 斑状积雪
patterned gravel structure 成型卵石结构
patterned ground 构造土,成型土
patterned ground forms 成型土类型
patterned ground origins 成型土成因
peak discharge（＝peak flow） 洪峰流量
peak time lag 洪峰滞时
peat 泥炭
peat bog 泥炭沼泽
peat deposit 泥炭沉积
peat soil 泥炭土
peatland 泥炭地
pedosphere 土壤圈,土壤层
Pee Dee Belemnite（PDB） （美国南卡罗来纳州白垩系皮狄组的）美洲拟箭石（PDB）国际碳同位素标准
pellicular water 薄膜水,附着水
penetration depth 渗透深度,穿透深度
penetration radiation 穿透辐射
penetrative convection 贯穿对流

penetrative downdraft 贯透下沉气流
penetrometer 硬度计
penguin 企鹅
penknife ice 刀〔状〕冰
pentad 候,逐候
perch pond 冰面湖
perched cirque 悬冰斗
perched dune 坡顶沙丘
percolation zone 渗浸带,渗透带
perennial 多年生的,长年的
perennial ice-cover 常年冰封
perennial snow patches 多年雪斑
perennial snowbank 多年雪堆
perennially frozen ground 多年冻土
perennially frozen lake 长年封冻湖
perennially frozen sediment 多年冻结沉积物
perfectly plastic material 理想塑性体
periglacial 冰缘的,冰缘区的
periglacial action 冰缘作用
periglacial area（＝periglacial region） 冰缘地区
periglacial climate 冰缘气候
periglacial deposit 冰缘沉积
periglacial environment 冰缘环境
periglacial geomorphology 冰缘地貌
periglacial gesteinfluren 冰缘稀疏植物区
periglacial hydrology 冰缘水文学
periglacial involution 冰缘扰动

periglacial lake　冰缘湖
periglacial landform　冰缘地形
periglacial landform spectrum　冰缘地貌序列
periglacial landforms　冰缘地貌
periglacial landscape　冰缘地形，冰缘景观
periglacial loess　冰缘黄土
periglacial manifestation　冰缘现象
periglacial morphogenetic　冰缘地貌形态的
periglacial phase　冰缘相
periglacial phenomenon　冰缘现象
periglacial processes　冰缘过程
periglacial sediment　冰缘沉积物
periglacial slopewash　冰缘坡面冲刷
periglacial stage　冰缘期
periglacial terrace　冰缘阶地
periglacial tor　冰缘岩柱
periglacial valley　冰缘河谷
periglacial zonation　冰缘〔分〕区〔带〕
periglacial zone　冰缘地带
periglacio-geomorphic feature　冰缘地貌特征
periglaciology　冰缘学
perihelion　近日点
permafrost　多年冻土
permafrost age　多年冻土年龄
permafrost aggradation　多年冻土扩张
permafrost base　多年冻土下限
permafrost belt　多年冻土带
permafrost boundary　多年冻土边界
permafrost condition　多年冻土状况
permafrost deep drilling　多年冻土深钻
permafrost degradation　多年冻土退化
permafrost distribution　多年冻土分布
permafrost division　多年冻土分区
permafrost dynamics　多年冻土动力学
permafrost engineering　多年冻土工程
permafrost feature　多年冻土特征
permafrost forecasting（＝permafrost prognose）　多年冻土预报
permafrost horizon　多年冻土层
permafrost hydrogeology　多年冻土水文地质学
permafrost layer　多年冻土层
permafrost limit　多年冻土界限
permafrost line survey　多年冻土路线勘查
permafrost map　多年冻土图
permafrost map unit　多年冻土制图单元
permafrost mapping　多年冻土制图
permafrost mapping technology　多

年冻土制图技术
permafrost pedon　多年冻土单个土体
permafrost polypedon　多年冻土聚合体
permafrost prediction　多年冻土预测
permafrost profile survey　多年冻土剖面调查
permafrost prognose（=permafrost forecasting）多年冻土预报
permafrost projection　多年冻土预估
permafrost region　多年冻土区
permafrost regionlization　多年冻土区划
permafrost southern boundary　多年冻土南界
permafrost survey　多年冻土调查
permafrost survey manual　多年冻土调查手册
permafrost table　多年冻土上限
permafrost taxonomy　多年冻土分类系统
permafrost temperature　多年冻土温度
permafrost thawing　多年冻土融化
permafrost thermal regime　多年冻土热状况
permafrost thickness　多年冻土厚度
permafrost type　多年冻土类型

permafrost zonation　多年冻土分带
permafrost zone　多年冻土带
permagelic gleyosols　多年冻土潜育土
permagelic histosols　多年冻土有机土
permanent community（=persistent community）稳定群落
permanent lake　常年湖泊
permanent pasture　稳定草场
permanent station　长期水文站
permanently frozen ground　多年冻土
permeable bed　透水层
permeable boundary　透水边界
permeable medium　可渗透介质
permic geli-alluuvic primosols　多年寒冻冲积新成土
permic geli-orthic primosols　多年寒冻正常新成土
permic geli-sandic primosols　多年寒冻砂质新成土
permittivity　渗透性
Peroxyacetyl Nitrate（PAN）硝酸过氧化乙酰
persistent organic pollutants（POPs）持久性有机污染物
perturbation　扰动
petrel　海燕
petrogypsic anhyorthels　岩质石膏脱水正常寒冻土
petrogypsic anhyturbels　岩质石膏

脱水扰动寒冻土
phase change 相变
phase transformation 相变过程
Phased Array type L-band Synthetic Aperture Radar（PALSAR） 相位型 L 波段合成孔径雷达
phosphate 磷酸盐
phosphorite 磷钙土
phosphorus 磷
phosphorus cycle 磷循环
photo interpretation 像片判读
photo mosaic 像片镶嵌
photoassimilation 光同化〔作用〕
photochemical smog 光化学烟雾
photoelectric sensor 光电传感器
photogrammetric coordinate system 摄影测量坐标系
photogrammetric distortion 摄影测量畸变
photogrammetry 摄影测量学
photographic image 摄影影像
photographic processing 摄影处理
photographic reconnaissance satellite 照相侦察卫星
photography 摄影学
photoinhibition 光抑制
photoionization 光电离
photomicrography 显微摄影
photon 光子
photoperiodism 光周期
photorespiration 光呼吸
photosynthate 光合作用产物
photosynthesis 光合作用
photosynthetically active photon flux density（PPFD） 光量子通量密度
photosynthetically active radiation（PAR） 光合有效辐射
phreatic zone 稳定饱和带
phtosynthetically active radiation 光合有效辐射
phyplankton 浮游植物
physical weathering 物理风化
physiographic stage 地文期
physiology 生理学
phytoedaphon 土壤微生物群落
phytogeocoenosis 植物地理群落
phytology 植物学
phytomass（＝plant biomass） 植物生物量
phytoplankton 浮游植物
pick （两头尖的)镐
pick axe （丁字形)镐
piedmont deposit 山麓沉积物
piedmont glacier 山麓冰川
piedmont snow patch 山麓雪斑
pile foundation 桩基础
pilotless plane 无人飞机
pilum （孢粉)基柱
Pinatubo volcano 皮纳图博火山（菲律宾）
pinching frost 严寒,严霜
pingo 冻胀丘,冰丘
pingo decay 冻胀丘消退
pingo ice 冻胀丘冰
pingo lake 冰丘湖

pingo remnant 冰丘残迹
pingo scar 冰丘遗迹
pioneer species 先锋种
pioneer stage 先锋阶段
pioneer vegetation 先锋植物
pipe survey 管道勘察
pipkrake (=needle ice) 冰针(瑞典语)
Planck's radiation formula 普朗克辐射公式
planetary boundary layer (PBL) 行星边界层
planetary permafrost 行星多年冻土
planimeter (=platometer) 求积仪,测面仪
planned adaptation 有计划适应
plansol 湿草原土
plant indicator 指示植物,植物指示
plant nutrition 植物营养物,植物营养〔学〕
plastic deformation 塑性变形
plastic frozen ground 塑性冻土
plastic index 塑性指数
plastic limit 塑限
plate 板块
plate boundary 板块边界
plate convergence 板块汇聚
plate divergence 板块离散
plate ice 板状冰
plateau 高原
platy soil structure 板状土壤构造

playa 干盐湖
Pleistocene periglacial 更新世冰缘
Pleistocene permafrost 更新世多年冻土
Pliocene 上新世,上新世的
plough (=plow) 刨蚀
plowshares (=foam crust) (积雪表面消融时的)波状雪面形态
pluck (冰川)拔蚀,拔削
pluvial 多雨的,雨期的,洪积的,洪水成的
pluvial fan facies 洪积相
podzoliation 灰化〔作用〕
pogonip 冻雾
point observation 定点观测
point rainfall 点雨量
Poisson's ratio of frozen soil 冻土泊松比
polar bear (*ursus maritimus*) 北极熊
polar cod (*boreogadus saida*) 极地鳕鱼
polar cold temperate zone flora 极地寒冷植物群
polar desert tundra 极地沙漠苔原
polar glacier 极地冰川
polar ice cap 极地冰帽
polar ice sheet 极地冰盖
polar invasion (=outbreak) 寒潮
polar mesospheric clouds (PMC) 极地中间层云
polar orbit 极轨道

polar outbreak 寒潮爆发
polar plant community 极地植物群落
polar stratospheric cloud（PSC） 极地平流层云
polar willow（*salix polaris*） 极柳
polar zone 极区
Polarimetric L-band Multibeam Radiometer（PLMR） 全极化L波段微波辐射计
polarity chron 极性期
polarity event 极性事件
polarization radar 极化雷达
polar-orbiting meteorological satellite 极轨气象卫星
polar-type glacier 极地型冰川
poleward current 极向海流
policy selection matrix（PSM） 政策选择矩阵
polished surface 磨光面
polishing （冰川）磨蚀
pollen 花粉
pollen spectrum （孢粉）花粉谱
pollen statistics 花粉分析
pollen stratigraphy 花粉地层学
polychlorinated biphenyls（PCBs） 多氯联苯
polycrystalline ice 多晶冰
polycyclic aromatic hydrocarbon（PAH） 多环芳烃
polygon structure 多边形结构
polygon trough 多边形沟槽
polygonal cracking 多边形开裂
polygonal ground 多边形土
polygonal peat plateau 多边形泥炭高地
polynya（=polynia） 冰间水域（海冰），冰间湖，冰隙
polythermal glacier 多温型冰川
poorly-bonded permafrost 弱胶结多年冻土
population growth 人口增长
population structure 人口结构
pore ice 孔隙冰
pore medium 孔隙介质
pore ratio 孔隙比
pore space 孔洞
pore water 孔隙水
pore water pressure 孔隙水压力
pore-fissure aquifer 孔隙－裂隙含水层
pores 孔隙
porosity 孔隙度
porous aquifer 孔隙含水层
porous ice 多孔冰，疏松冰，多孔冰晶
porous medium 多孔介质
portable cup anemometer 便携式风杯风速表
portable inductive salinometer 便携式感应盐度仪
portable sampler 便携式采样器
porus （孢粉）孔
post-depositional process 后沉积过程
post-glacial 冰后期

post-glacial climate 冰后期气候
post-glacial climatic optimum 冰后期气候适宜期
post-glacial deposition 冰后期沉积
post-glacial period (=post-glacial age) 冰后期
post-glacial rebound 冰后期反弹
Post-Pliocene 晚上新世
potassium 钾
Potassium-Argon (K/Ar) Dating 钾－氩测年
potential energy 潜能
potential equivalent evaporation 潜在等效蒸发
potential evaporation 潜在蒸发
potential evaporation rate 潜在蒸发率
potential evapotranspiration 潜在蒸散发
potential flow 位(势)流
potential flow theory 势流理论
potential gradient 位势梯度
potential instability 位势不稳定
potential instability index 位势不稳定函数
potential melting capacity 潜在消融量
potential recharge 补给容量,持水能力
potential temperature 位温
potential transpiration 潜在蒸腾
potential visibility 有效能见度

potential vorticity 位势涡度
pothole 壶穴,锅穴,瓯穴
pothole erosion 壶穴侵蚀
potometer (=transpirometer) 蒸腾计
powdery snow (=powder snow) 粉末〔状〕雪,干粒雪
Power Reflection Coefficient (PRC) 功率反射系数
power spectrum 功率谱,能谱
pradoliny 冰蚀宽谷
prairie 草原,草地
prairie climate 草原气候
prairie community 草原群落
prairie snow 草原积雪
pratensis saffron (=saffron crocus=crocus sativus=prairie crocus) 藏红花
preboreal 前北方期
precautionary principle (PP) 预防原则
precession 岁差,进动
precession effect 岁差效应
precessional cycle 岁差周期
precipitable water 可降水量
precipitable water index 可降水量指数
precipitation 降水,降水量
precipitation acidity 降水酸度
precipitation aloft 高空降水
precipitation attenuation 降水衰减
precipitation band 降水带
precipitation cell 降水中心

precipitation chart　降水量图
precipitation chemistry　降水化学
precipitation collector　降水收集器
precipitation cumulus model　降水积云模式
precipitation detector　降水检测器
precipitation duration　降水持续时间
precipitation echo　降水回波
precipitation effectiveness　降水有效性
precipitation effectiveness ratio　降水有效比率
precipitation efficiency　降水效率
precipitation electricity　降水电学
precipitation flux　降水通量
precipitation gauge　雨量筒
precipitation generating element　降水生成胞，降水生成单体
precipitation index　降水指数
precipitation intensity　降水强度
precipitation particle spectrometer（PPS）　降水粒子谱仪
precipitation physics　降水物理学
precipitation process　降水过程
precipitation recorder　降水记录器
precipitation regime　降水情势
precipitation scavenging　降水清除
precipitation spectrometer probe（PSP）　降水粒子谱仪探头
precipitation stimulation　人工降水
precipitation surge　降水骤增
precipitation trajectory　降水轨迹
precipitation variation　降水变化
precipitation within sight　视区内降水
precipitation-effectiveness（P-E）index　有效降水指数
precipitation-evaporation（P-E）index　降水蒸发指数
precise traversing　精密导线测量
pre-cold frontal squall-line　冷锋前飑线
preconstruction thaw method　预融法
predatory exploitation　掠夺式开发
predictability　可预报性
predictand　预报量
predicting ice　冰情预报
prediction　预测
preglacial climate　冰缘气候
preglacial stream　冰缘水流
preglacial terrace　冰缘阶地
prehistoric period　史前期
Preliminaire EOLE（PEOLE）　"佩奥利"气象卫星
pre-processing　预处理
present landuse map　土地利用现状图
pressure altitude　气压高度
pressure anomaly　气压距平
pressure center　气压中心
pressure change　气压变化
pressure chart　气压图

pressure coefficient 压力系数
pressure coordinate system 气压坐标系
pressure distribution 气压分布
pressure drift sheet 挤压冰碛层
pressure equation 气压方程
pressure field 气压场
pressure gradient 气压梯度
pressure head 压力水头,压强水头
pressure ice (=screw ice) 挤压冰
pressure melt 压力融化
pressure melting point 压力融点
pressure profile 气压廓线
pressure recorder 气压记录器
pressure reduction 气压换算
pressure ridge 压力脊
pressure spectrum 气压谱
pressure surge 气压涌升
pressure surge line 气压涌升线
pressure system 气压坐标系
pressure trough 气压槽
pressure velocity 气压〔铅直〕速度
pressure vertical coordinates 气压垂直坐标
pressure-change chart 气压变化图
pressure-fall centre 降压中心
pressure-ice foot 挤压冰壁
pressure-melting temperature 压力融化温度
pressure-temperature correlation 压温相关
pressurized 加压的,受压的
prethawing method 预融法

prevailing westerlies 盛行西风带
prevailing wind direction 盛行风向
primary community 原生群落
primary ice 初生冰
primary phytocoenosium 原生植物群落
primary pollutant 原生污染物,一次污染物
primary pollution 一次污染
Princess Elizabeth Land 伊丽莎白公主地(南极)
principal shear strain 主剪应变
principal shear stress 主剪应力
principal strain 主应变
principal stress 主应力
principle component analysis (PCA) 主成分分析
probability density function (PDF) 概率密度函数
probable maximum precipitation (PMP) 可能最大降水
profiler 断面仪
proglacial 冰前的
proglacial delta 冰前三角洲
proglacial deposit 冰前沉积物,冰前沉积
proglacial lake 冰前湖
proglacial landform 冰前地形
proglacial meltwater 冰前融水
proglacial outwash (冰川前缘)冰水沉积
proglacial processes 冰前过程
proglacial river (=proglacial stream)

冰前河流
proglacial stream（＝proglacial river）　冰前河流
prognostic equation　预报方程
prognostic map　预报图
progressive wave　前进波
prokaryotes　原核生物
promojna　冰面开裂
proportional error　比例误差
proportional weir　比例分流堰
prosere　早期演替系列
prospecting baseline　勘探基线
prospecting line survey　勘探线测量
prospecting network layout　勘探网测设
protalus rampart　倒石堆前缘堤
protozoa　原生动物
provincialism　地区性
proxy climate indicator　气候代用指标
proxy climate record　代用气候记录
proxy data　代用资料
psammentic aquiturbels　粗粒含水扰动寒冻土
psammentic aquorthels　粗粒含水正常寒冻土
psammoeremion　荒漠群系
psammorthels　粗粒正常寒冻土
psammoturbels　粗粒扰动寒冻土
pseudo cold-front　假冷锋
pseudo front　假锋
pseudo-adiabat　假绝热

pseudo-color image　假彩色图像
pseudo-wet-bulb temperature　假湿球温度
Psychozoic Era　灵生代
psychrograph　干湿计
psychrometer　干湿表,玻璃水银干湿表
psychrophile　嗜冷菌
psychrophile（＝psychrophilic organism）　嗜冷生物,低温微生物
psychrophilic　喜低温的,嗜冷的
psychrophilic microorganism　嗜冷微生物
psychrophyte　高山寒土植物
psychrotolerant microorganism　耐冷微生物
psychrotroph　耐冷菌
ptenophyllium　落叶林群落
ptenophyllophyte（＝ptenophyllophyta）　落叶植物
pteris　凤尾蕨属
pterpod ooze　翼足虫软泥
public adaptation　公共适应
puck　冰球
pulk　（单鹿拉的）船形雪橇
pull fringing coast　拖曳边缘海岸
pulsed source　脉冲源
Pumi Nationality　普米族
pure air　纯空气
pure gravity anomaly　纯重力异常
pure ice　纯冰
pure polycrystalline ice　纯净多晶冰

pure shear 纯剪切
purification 纯化
push moraine 推出碛
push-broom imaging 推扫式成像
pycnocline 密度跃层
pyramid dune 金字塔沙丘
pyramid peak（=glacier horn） 角峰
pyroclast 火山碎屑物

Q

qanat（=karez） 坎儿井，暗渠
Qiang Nationality 羌族
Qinghai-Xizang（Tibet）Highway 青藏公路
Qinghai-Xizang（Tibet）movement 青藏运动
Qinghai-Xizang（Tibet）Plateau 青藏高原
Qinghai-Xizang（Tibet）Plateau monsoon 青藏高原季风
Qinghai-Xizang（Tibet）Railway 青藏铁路
Qinghai-Xizang（Tibet）trough 青藏低槽
Qomolangma Glaciation 珠穆朗玛冰期
Qomolangma National Nature Preserve（QNNP） 珠穆朗玛峰国家级自然保护区
qualitative approach 定性评估方法
quantitative analysis 定量分析
quantitative forecasting 定量预报
quantitative spectrometric analysis 光谱定量分析
Quar Ice Shelf 奎尔冰架(南极)
quarried surface 挖掘面
quarrying 挖掘作用
quartz 石英
quartz thermometer 石英温度计
quasi-equilibrium 准平衡
quasi-nondivergence 准无辐散
quasi-oscillation 准周期振动
quasi-periodic 准周期性
quasi-periodicity 准周期
Quaternary 第四纪
Quaternary climate 第四纪气候
Quaternary climatic record 第四纪气候记录
Quaternary deposit 第四纪沉积
Quaternary eustatic movement 第四纪海平面升降运动
Quaternary fold 第四纪褶皱
Quaternary geology 第四纪地质学
Quaternary glacial period 第四纪冰期
Quaternary glaciation 第四纪冰川作用

Quaternary Ice Age 第四纪大冰期
Quaternary lacustrine soil 第四纪湖相母质土壤
Quaternary ore deposit 第四纪矿床
Quaternary period 第四纪
Quaternary red earth 第四纪红土
Quaternary sediment 第四纪沉积物
Quaternary System 第四系
QuickBird 快鸟卫星(美国)
quiescence 静止
quiescent period 静止期

R

radar 雷达
radar algorithm 雷达算法
radar altimeter 雷达高度计
radar altimetry 雷达测高学,雷达测高术
radar antenna 雷达天线
radar band 雷达波段
radar beam 雷达波束
radar calibration 雷达校准
radar clutter 雷达杂波
radar coordinate 雷达坐标系
radar coverage 雷达覆盖范围
radar data interpretation 雷达数据解译
radar digital display 雷达数字显示〔器〕
radar glaciology 雷达冰川学
radar hail detector 雷达冰雹探测器
radar interferometry 雷达干涉测量
radar layer 雷达层位
radar longitudinal profile 雷达纵剖面
radar map 雷达图
radar mast 雷达天线杆
radar mosaic 雷达拼图
radar navigation 雷达导航
radar profile 雷达廓线
radar ramark 雷达指向标
radar reflection 雷达反射
radar reflection interval 雷达反射时间间隔
radar reflectivity 雷达反射率
radar remote sensing 雷达遥感
radar resolution 雷达分辨率
radar responder 雷达应答器
radar return 雷达回波
Radar Satellite (= Radarsat) 雷达卫星(加拿大)
radar scanning 雷达扫描
radar section 雷达剖面

radar shadow 雷达盲区
radar sounder 雷达探测器
radar sounding 雷达探测
radar stereo-viewing 雷达立体观测
radar stratigraphy 雷达层位〔剖面〕
radar survey 雷达测量
radar target 雷达目标
radar transmitter 雷达发射机
radar transparency 雷达透视
radar wave attenuation 雷达波衰减
radar wave energy 雷达波能量
radar wave frequency 雷达波频率
radar wave length 雷达波长
radar wave refraction 雷达波折射
radar wave velocity 雷达波速
radargram 雷达图像
radargrammetry 雷达测量,雷达摄影测量
radar-probing system 雷达探测系统
Radarsat（＝Radar Satellite） 雷达卫星(加拿大)
radarscope 雷达示波器
radial crevasse 放射状冰隙
radial faults 辐射状断层
radial triangulation 辐射三角测量
radiance 辐射率
radiating glacier 放射状冰川
radiation 辐射

radiation balance 辐射平衡
radiation balance equation 辐射平衡方程
radiation budget 辐射收支
radiation cooling 辐射冷却
radiation energy 辐射能
radiation exitance 辐射度
radiation feedback process 辐射反馈过程
radiation flux 辐射通量
radiation flux density 辐射通量密度
radiation frost 辐射霜冻
radiation frost injury 辐射霜冻害
radiation intensity 辐射强度
radiation inversion 辐射逆温
radiation loss 辐射亏损
radiation melt 辐射消融
radiation pattern avalanche 辐射状雪崩
radiation sensor 辐射传感器,辐射探头
radiation source 辐射源
radiation spectrum curve 辐射光谱曲线
radiation spectrum function 辐射光谱函数
radiation transfer 辐射传输
radiation transfer model 辐射传输模型
radiative equilibrium 辐射平衡
radiative equilibrium temperature 辐射平衡温度

radiative forcing　辐射强迫
radical scavenger　自由基清除剂
radio altimeter　无线电测高计
radio echo-sounding（RES）　无线电回波探测
radioactive　放射性的
radioactive aerosol　放射性气溶胶
radioactive carbon（=radiocarbon）　放射性碳
radioactive dating　放射性定年
radioactive decay　放射性衰变
radioactive deposit　放射性沉积
radioactive element　放射性元素
radioactive fallout　放射性沉降
radioactive horizon　放射性层位
radioactive isotope　放射性同位素
radioactive snow gauge　放射探测自动雪量计
radioactive tracer　放射性示踪物
radioactivity　放射性
radiocarbon　放射性碳
radiocarbon chronology　放射性碳年代学
radiocarbon dating　放射性碳定年
radiocarbon geochronology　放射性碳地质年代学
radio-echo record　无线电回波记录
radio-echo sounding　无线电回波探测
radioglaciology　无线电冰川学
radiometric　放射性测量的，辐射度的
radiometric age　放射性年龄
radiometric calibration　辐射订正，辐射标定
radiometric correction　辐射校正
radiometric dating　放射性定年
radiometric resolution　辐射分辨率
radionuclide　放射性核素
radio-wave velocity　无线电波速
rafted boulder　冰川漂砾
rafted ice　筏冰,重叠冰
rafted ice（=telescoped ice）　叠加海冰
rafted ice thickness　重叠冰厚度
rafting　（海冰）叠加,叠层
railway switch deiceing　铁路道叉除冰
rain crust　雨斑壳,雨结壳
raindrop freezing　雨滴冻结
rainfall excess　产流雨量
rainfall runoff　降雨径流
ram penetrometer　冰雪硬度计,冰雪穿透计
ram sonde　冰雪硬度器
rams　冰角
random distribution（=random dispersion）　随机分布
random sampling　随机取样
range management（=pasture management）　牧场管理
rangeland ecology（=range ecology）　牧场生态学
range-only radar　测距雷达

ranging instrument 测距仪
rare earth element（REE） 稀土元素
rare species 稀有种
raster data 栅格数据
raster format 栅格格式
ratio enhancement 比值增强
rational grazing 定额放牧
raw data 原始数据
Rayleigh fractionation 瑞利分馏
Reyleigh model 瑞利模型
Reyleigh scattering 瑞利散射
real-time ice particle size analysis system 冰粒尺度实时分析系统
real-time photogrammetry 实时摄影测量
real-time processing 实时处理
reanalysis data 再分析资料
receding glacier 退缩冰川
recemented glacier（＝glacier remani＝reconstituted glacier＝regenerated glacier） 再生冰川
recession curve 退水曲线
recession limb （过程线的）退水段
recessional moraine 冰退冰碛垄
recessional outwash 冰退冰水沉积
reconnaissance camera 勘测相机
reconstituted glacier（＝glacier remani＝recemented glacier＝regenerated glacier） 再生冰川
reconstruction 重建,恢复
recording albedo meter 反照率自记计
recrystallization 重结晶作用
recrystallization zone 重结晶带
refilling interval （冰湖）蓄水期
reflectance 反射率
reflectance spectrum 反射波谱
reflecting layer 反射层
reflection angle 反射角
reflection coefficient 反射系数
reflection intensity 反射强度
reflection nephoscope 反射测云器
reflection sounding 回声探测
refracted wave 折射波
refraction angle 折射角
refraction index 折射指数
refreezing 再冻结
refrigerant 制冷剂
refrigerator 制冷机
refugee 难民
regelation 复冰作用
regelation deformation 再结冰变形
regelation layer 复冰层
regenerated glacier（＝glacier remani＝recemented glacier＝reconstituted glacier） 再生冰川
regional climate 区域气候
regional climate model（RegCM） 区域气候模式
regolith 风化层
regosolic static cryosol 岩成静态寒冻土
regosolic turbic cryosol 岩成扰动

寒冻土
regression analysis 回归分析
regular reflector 镜面反射体
reindeer（*rangifer tarandus*） 北极驯鹿
relative humidity 相对湿度
relative ice content 相对含冰量
relative permittivity 相对电容率
relic ice（=relict ice） 残留冰
relict fauna 残遗动物区系
relict flora 残遗植物区系
relict flow stripe 残遗冰流带
relict glacier 残余冰川
relict lake 残遗湖
relict permafrost 残遗多年冻土
relict permafrost table 残遗多年冻土上限
relict pingo 残遗冰皋,残遗冻胀丘
relict sediment 残余沉积物
relict thermokarst 残遗热融喀斯特
remote sensing 遥感
remote sensing image 遥感影像
remote sensing platform 遥感平台
remote sensing sounding 遥感测深
remote sensor 遥测传感器
repose angle 休止角
representative concentration pathways（RCP） （温室气体排放的）典型浓度路径
resampling 再抽样,重采样
residence time 滞留时间
residual border ice 残余岸冰
residual ice accumulation 残冰堆积
residual material 残积物
residual strength 残余强度
residual stress 残余应力
residual thaw layer 残余融化层,残留融层
resistance coefficient 阻力系数
respiration 呼吸作用
respiration equation 呼吸方程
response function 响应函数
response lag 响应滞后
response time 响应时间
rest-rotation grazing 休闲轮牧
retention ability 持水力
retention factor 持水系数
retention time 滞水时间
reticulate （孢粉）网状雕纹
reticulate ice 网状冰
reticulate-blocky cryostructure 格状－块状冷生结构
retreatal moraine 后退冰碛
retrogressive evolution 后退演化
retrogressive thaw slump 后退型（溯源）热融滑塌
Reynolds equation 雷诺方程
Reynolds number 雷诺数
Reynolds stress 雷诺应力
rheidity 流变性
rheology 流变,流变学
rhinoceros tichorhinus 披毛犀
rhythmic bedding 韵律层理

rhythmite 带状纹泥层（冰川湖中的沉积层）
ribbon ice 条纹冰
Richardson number 里查森数
richness index 丰富度指数
ridged ice zone 冰脊带
riegel 冰坎,冰谷横阶
rift swarm 冰裂谷群
rift valley 裂谷
rigidity 硬度,刚性
Riiser-Larsen Ice Shelf 里塞一拉森冰架（南极）
rimaye（=bergschrund） 粒雪盆后壁裂隙
rime 雾凇,不透明冰
rime fog 霜雾,冰雾
rime ice 雾凇冰,冰凇
rime rod 雾凇探棒
rime sensor 雾凇探测器
rime-splinter process 凇碎过程
rind ice （河、湖、海）壳冰
rink 冰球场
rinnental 冰壁沟
rinnentaler 碛堤河槽
ripe snow 老雪
risk assessment (RA) 风险评估
risk management (RM) 风险管理
risk-prone region 风险易发区
river ice 河冰
river ice break-up 开河
river ice jam 河冰壅塞
river ice run 淌凌
river ice-drifting 河流流冰,河流淌凌
river icing conditions 河流冰情
river talik 河流融区
roadbed cooling 冷却路基
roaring forties 咆哮西风带（南纬40°～50°风暴带）
robin snow 春天小雪
roche moutonnee 羊背石
rock debris 岩屑
rock drumlin 石鼓丘
rock field 石海
rock glacier 石冰川
rock glacier front 石冰川前缘
rock glacier meltwater 石冰川融水
rock piton 岩石锥
rock stream 石河
rockfall talus 落石型倒石堆
Rocky Mountains 落基山
rofla 冰河谷（上游宽下游窄,常见于瑞士）
rolling road surface （冻融作用导致的）起伏路面
Ronne Ice Shelf 罗尼冰架（南极）
roof snow loads 房顶雪载荷
rope ladder（=tape ladder） 小挂绳梯
ross seal（*ommatophoca rossii*） 罗斯海豹
Rossby number 罗斯贝数（大气）
Rossby wave 罗斯贝波（大气）
rotary meter 流速仪
rotation cycle 轮作周期

rotation grazing 轮牧
rotation pasture 轮牧地
rotten ice 蜂窝状冰
rotten stone 蜂窝石
rough ice 粗糙冰
roughness 粗糙度
roughness parameter 粗糙度参数
roughness spectral function 粗糙度谱函数
royal penguin (*eudyptes schlegeli*) 皇企鹅
rubber ice 松冰团,弹性冰
rubble ice 碎块冰
rundhall 羊背石,鼓丘地形
running water sediment 流水沉积物
runoff 径流
runoff coefficient 径流系数
runoff erosion 径流侵蚀
runoff modulus (= flow modulus) 径流模数
runoff variation 径流变化
runoff yield 产流
ruptic historthels 不规则有机正常寒冻土
ruptic histoturbels 不规则有机扰动寒冻土
ruptic-histic aquiturbels 不规则有机含水扰动寒冻土
ruptic-histic aquorthels 不规则有机含水正常寒冻土
Russian Nationality 俄罗斯族

S

Saddle Glacier 赛德勒冰川(加拿大)
saddle pebble 马鞍石
sag subduction 沉陷俯冲
sailing direction 航路指南,航路指示
sailing ice 稀疏流冰,漂凌
salic anhyorthels 盐积脱水正常寒冻土
salic anhyturbels 盐积脱水扰动寒冻土
salic aquorthels 盐积含水正常寒冻土
salic crust 盐结表层
salic evidence 盐积现象
salic horizon 盐积层
salination (= salinization) 盐渍化
saline frozen soil 盐渍化冻土
saline permafrost 盐渍多年冻土
salinity 盐度
salinity index 盐度指数
salipan horizon 盐磐
Salpausselk 萨尔珀冰碛丘陵,后退冰碛(芬兰)

salt corrie 冰斗状盐凹地
salt desert 盐漠
salt flower（＝ice flower） 冰花，盐花（海水快速冻结所形成的霜晶）
salt glacier 盐冰川（盐栓呈舌形在坡上伸下）
salt lake 咸水湖
sample area 取样区，样本区
sample bottle 样品瓶
sample capacity 样本容量
sample interval 采样间隔
sample plot 取样点，样本点
sample region 样区
sample size 样本大小
sample standard deviation 样本标准差
sample survey 抽样调查
sample value 样本值
sample variance 样本方差
sampling 采样
sampling error 采样误差
sampling flask 采样瓶
sampling frequency 采样频率
sampling inspection 抽样检查法
sampling protocol 采样方案
sampling site 采样点
sampling system 采样系统
sampling unit 抽样单元
sand bar 沙坝
sand barrier 沙障
sand fixing afforestation 固沙造林
sand snow 沙状雪
sand wedge 砂楔
sand wedge casts 砂楔假型
sand-fixation forest 固沙林
sandstone 砂岩
sandstorm 沙尘暴
sandur 冰水沉积平原
sand-wedge polygon 砂楔多边形
sandy soil 砂土
sapric fibri-permagelic histosols 高腐纤维多年冻结有机土
sapric glacistels 高腐厚冰有机寒冻土
sapric hemi-permagelic histosols 高腐半腐多年冻结有机土
sapric soil materials 高腐有机土壤物质
sapristels 高腐有机寒冻土
sapropel 腐泥，腐殖泥
sapropelic 腐泥的
saprophyte（saprobe） 腐生植物
saprozoic 食腐动物－腐生的，腐生植物的
sastrugi（skavl） 雪面垄坎，雪垄（俄语）
satellite altimetry 卫星测高
satellite altitude 卫星高度
satellite attitude 卫星姿态
satellite climatology 卫星遥感气候学
satellite cloud photograph 卫星云图
satellite coverage 卫星覆盖范围

satellite Doppler positioning　卫星多普勒定位
satellite geodesy　卫星大地测量学
satellite gradiometry　卫星重力梯度测量
satellite image　卫星影像
satellite infrared spectrometer　卫星红外光谱仪
satellite interpretation message（SIM）　卫星解译信息
satellite laser radar　卫星激光雷达
Satellite Laser Ranging（SLR）　卫星激光测距法
satellite meteorology　卫星气象学
satellite monitroing　卫星监测
satellite oceanography　卫星海洋学
satellite orbit　卫星轨道
satellite photo map　卫星像片图
satellite positioning　卫星定位
Satellite Pour l'Observation de la Terre（SPOT）　斯波特卫星（法国）
satellite remote sensing　卫星遥感
satellite sounding　卫星探测
satellite tracking station　卫星跟踪站
satellite-borne sensor（= space-borne sensor）　星载传感器
satellite-borne weather radar　星载气象雷达
Satellite-to-Satellite Tracking（SST）　星间定位技术

satin ice（= acicular ice）　针冰，冰针
saturability（= saturation degree）　饱和度
saturated adiabatic　饱和绝热
saturated adiabatic lapse rate（SALR）　饱和绝热温度递减率
saturated air　饱和空气
saturated pereability　饱和透水性
saturated vapour　饱和水汽
saturated virtual temperature　饱和虚温
saturated zone　饱和区,饱和带
saturation　饱和度
saturation curve　饱和曲线
saturation deficit　饱和差
saturation dose　饱和剂量
saturation point　饱和点
saturation vapor pressure　饱和水汽压
savannas　萨瓦纳草原,热带稀树草原
scale characteristics　尺度特征
scale effect　尺度效应
scale model　比例模型(模具)
scale parameter　尺度参数
Scandinavia　斯堪的纳维亚
Scandinavian Ice Sheet　斯堪的纳维亚冰盖（古）
Scanning Hydrographic Operational Airborne Lidar Survey（SHOALS）　扫描式水道测量机载激光雷达系统

scanning microwave radiometer 扫描式微波辐射计
scanning microwave spectrometer (SCAMS) 扫描式微波频谱仪
Scanning Multichannel Microwave Radiometer (SMMR) 多通道微波扫描辐射计
scattering body 散射体
scattering coefficient 散射系数
scattering cross section 散射截面
scattering light 散射光
scattering phase function 散射相函数
scattering radiation 散射辐射
scattering strength 散射强度
schmutzband 污化层
schneebrett 雪崩（德语）
schrund line 古冰斗线
scooped lake 侵蚀湖
scoring 冰擦作用，冰川擦痕
Scott Base 斯科特基地（南极，新西兰站）
scour 洼地，冲蚀洼地
scouring velocity 冲刷速度
scratch 擦痕，划痕
scree 岩屑（堆）
screw ice 挤压冰（河、湖、海冰）
screwing 旋转冰块壅塞，浮冰漩涡
sculpture （孢粉）雕纹，刻蚀
sea ice 海冰
sea ice characteristics 海冰特征
sea ice concentration 海冰密集度
sea ice condition 海冰冰情
sea ice detection 海冰检测
sea ice disaster 海冰灾害
sea ice dynamics 海冰动力学
sea ice engineering 海冰工程
sea ice environment 海冰环境
sea ice forecast 海冰预报
sea ice growth 海冰生长
sea ice mechanics 海冰力学
sea ice model 海冰模型
sea ice monitoring 海冰监测
sea ice motion 海冰运动
sea ice parameter 海冰参数
sea ice prediction 海冰预测
sea ice strength 海冰强度
sea ice temperature 海冰温度
sea ice thickness 海冰厚度
sea ice zone 海冰区
sea level correction 海平面订正
sea level pressure 海平面气压
sea salt 海盐
Sea Satellite (Seasat) （美国国家航空航天局）洋面风场探测卫星
sea surface albedo 海面反照率
sea surface height 海面高度
sea surface roughness 海面粗糙度
sea surface temperature (SST) 海面温度
sea surface topography 海面地形
sea temperature 海温
sea water freezing point 海水冰点
sea water intrusion 海水入侵
sea-ice chlorinity 海冰氯度

sea-ice frequency 海冰频率
seal (*arctocephalinae phocidae*) 海豹
sea-land breeze 海陆风
sea-land water cycle 海陆水循环
sea-level change 海平面变化
seanonal ice zone 季节性冰带
Seasat-A Satellite Scatterometer (SASS) 海洋卫星散射仪
seasonal creep 季节性蠕变
seasonal freezing 季节性冻结
seasonal freezing depth 季节冻结深度
seasonal freezing index 季节冻结指数
seasonal freezing layer 季节冻结层
seasonal frost 季节冻结
seasonal frost cracks 季节性冻胀裂缝
seasonal grazing 季节性放牧
seasonal lake 季节湖
seasonal snow cover 季节性积雪
seasonal thawed layer 季节融化层
seasonal thawing 季节融化
seasonal thawing index 季节融化指数
seasonal variation 季节变化
seasonality 季节性
seasonally frozen ground 季节冻土
seasonally frozen layer 季节冻结层
seasonally frozen soil 季节冻土
seasonally thawed ground 季节融土
seasonally thawed layer 季节融化层
seasonally thawed soil 季节融土
sea-state scale 海况标尺
seawater thermometer 海水温度表
secondary forest 次生林
secondary glacier 支冰川
secondary particle 次生粒子（宇宙）
secondary succession 次生演替
second-year ice 隔年冰
secular drift 缓慢漂移
sedentary farming 定居耕种
sedentary grazing 定居放牧
sedimentary basin 沉积盆地
sedimentary cycle 沉积旋回
sedimentary environment 沉积环境
sedimentary rhythm 沉积韵律
sedimentary structure 沉积构造
sedimentation 沉积作用
sedimentology 沉积学
seepage line 渗流线
seepage velocity 渗流速度,渗透速度
seesaw structure 跷跷板现象,跷跷板结构
segregated ice 分凝冰
segregation potential 分凝势
seismic sounding 地震探测〔法〕
self-restoration 天然更新
semi-arid 半干旱
semi-confined aquifer 半承压含水层
semi-intensive grazing system 半

集约式放牧制
semilogarithmic scale 半对数标度
sensible heat 感热
sensitivity analysis 敏感性分析
sensitivity test 敏感性实验
sensor network 传感器网络
sensor technique 传感器技术
septarian nodule 龟背条痕石
sequence stratigraphy 序列地层学
serac （冰裂隙之间的）脊状冰
settled snow （高密度）陈雪
Shackleton Ice Shelf 沙克尔顿冰架
shade tolerance (＝shade endurance) 耐阴性
shading board embankment 遮阳板路基
shadow band pyranometer 遮光罩天空辐射计
shale ice 薄碎冰群（河、湖）
shallow ice approximation (SIA) 浅冰近似〔理论〕
shallow ice core 浅冰芯
shape factor 形状因子
shear band 剪切带
shear deformation 剪切变形
shear failure load 剪切破坏荷载
shear fault 剪断层
shear fracture 剪切断裂
shear line 剪切线
shear modulus 剪切模量，剪变模量
shear resistance 剪切阻力
shear ridge 剪切冰脊（海冰）

shear ridge field 剪切冰脊区（海冰）
shear strain rate 剪切应变率
shear strength 剪切强度
shear stress 剪切应力
shear surface 剪切面
shear wave velocities 剪切波速
shear zone 剪切带
shearing deformation 剪切变形
shearing vorticity 切变涡度
sheepback rock (＝roche moutonnee) 羊背石
sheet flow 片流,薄层水流
sheet frost 片霜
sheet ice 片状冰,表面冰
sheet structure 层状构造
shelter 避难所,遮盖物
Sherpa 夏尔巴人,舍巴人
shielding effect 屏蔽效应,遮阳效应
shifting snow dune （风吹雪形成的）雪丘
Shipboard Ice Alert and Monitoring System (SAUS) 船载冰情报警与监控系统
shoal 浅滩
shore ice 岸冰
shore ice ride-up 岸冰上爬,海冰上岸过程
shore lead 沿岸冰间水道
shore polynya 沿岸冰间湖
short-term frozen ground 短时冻土

shortwave radiation 短波辐射
shoved moraine 推出碛
shovel 铁锹
shower 阵雨
showery snow 阵雪
shrub forest 灌丛林
shrubland 灌丛带
shuga （海水中的）白松冰团
Shuttle Imaging Radar（SIR） 航天飞机成像雷达
Shuttle Radar Topography Mission（SRTM） 航天地形测绘任务
Siberian husky 西伯利亚雪橇犬
side intersection 侧方交会
Side-Looking Airborne Radar（SLAR） 机载侧视雷达
side-looking radar（SLR） 侧视雷达
sighting distance 视距
sightseeing district 游览区
signal attenuation 信号衰减
signal intensity 信号强度
signal lamp 信号灯
signal pole 信号杆
signal-to-noise ratio（SNR） 信噪比
sikussak 峡湾老冰
silt 粉沙
similarity theory 相似理论
simple shear 单剪切
simulation 模拟
simulation condition 模拟条件
simultaneity 同时性
single crystal ice 单晶冰
single look complex（SLC） 单视复数据
single-aliquot regenerative-dose method（SAR method） 单片再生剂量法
single-pole COF（＝crystal orientation fabric） 单极型冰晶组构
Sinian Period 震旦纪
sintering 烧结作用
sinuosity 曲率，弯曲度
siphon rain-gauge 虹吸式雨量计
skare 雪壳〔古因纽特语〕
skating 滑冰
skavl（sastrugi） 雪面垄坎，雪垄（挪威语）
ski 滑雪
ski area 滑雪区
ski equipment 滑雪装备
ski helm 滑雪头盔
Ski Jumping 跳台滑雪（滑雪运动项目之一）
ski lift 滑雪缆车
ski resort 滑雪场
ski runway 滑雪道
ski season 滑雪季节
ski stick 雪杖
skiing 滑雪
skim ice（＝ice rind） 薄冰壳
skin temperature 表皮温度（遥感）
skiplane 雪上飞机
ski-scooter/skidoo 雪地摩托
skuas 贼鸥
sky burial（＝celestial burial） 天葬

sky radiation　天空辐射
sky radiometer（＝diffusometer）
　天空辐射表
slack ice　缓流冰〔屑〕
slade　冰斗状小穴－平地
slalom　障碍滑雪
slant range　斜距
sled runner　雪橇滑行板
sleeping bag cover　睡袋套
sleet　雨夹雪
sleigh　雪橇
sliding rate　滑动速率
sliding resistance　滑动阻力
sliding velocity　滑动速度
slob ice　海面破碎冰
slumping　滑塌
slush（＝soaked snow）　雪浆
slush avalanche　湿雪崩
slush ice　冰浆
slush ice sluice　泄冰闸
slush zone　雪浆带
slush-agglomerated ice　冻结冰花
small cake ice　小冰块（直径小于 2 m）
small ice floe　小冰盘,小块浮冰（宽度 9～180 m）
smog　烟雾
smoothing filter　平滑滤波器
snorter　强风,暴风雪
snow　雪
snow ablation　积雪消融
snow absorptivity　雪吸收率
snow accumulation　积雪累积

snow age　雪龄
snow albedo　积雪反照率
snow albedo feedback　雪反照率反馈〔机制〕
snow algae（*chlamydomonas nivalis*）　雪衣藻
snow amount　积雪量
snow anthor　雪锥
snow bank　雪堤
snow banner（＝snow plume＝snow smoke）　雪旗
snow barchan（＝snow medano）　新月形雪丘
snow belt　雪带
snow bin　雪量箱,量雪器,量雪箱
snow blindness　雪盲〔症〕
snow blink（＝snow sky）　雪映白光
snow blizzard　暴风雪
snow blower（＝snow thrower）　扫雪机,螺旋桨式除雪机
snow board　测雪板,量雪板
snow bridge　雪桥
snow capped peak　雪峰
snow chemistry　雪化学
snow classification　雪分类
snow climate（＝polar climate）　极地气候,雪原气候
snow cloud　雪云
snow coach　雪地汽车
snow concrete　硬雪,坚雪
snow cornice　雪堆
snow course　测雪路线
snow cover　积雪

snow cover area (SCA) 积雪面积
snow cover chart 积雪图
snow cover distribution 积雪分布
snow cover extent 积雪范围
snow cover fraction (SCF) 积雪覆盖率(度)
snow cover mapping 积雪制图
snow cover model 积雪模型
snow crust 雪壳,薄冰壳
snow crystal 雪晶
snow crystal concentration meter 雪晶浓度计(气象)
snow crystal size 雪晶尺寸
snow data assimilation system (SNODAS) 积雪数据同化系统
snow day 雪日
snow density 雪密度
snow depletion model 积雪衰退模型
snow deposit 雪沉积
snow depth 雪深
snow depth meter 雪深计
snow depth probe 雪深探头
snow depth scale 雪尺
snow diagenesis 雪成冰作用
snow disaster 雪灾
snow disaster forecast 雪灾预报
snow drift 吹雪堆,风吹雪
snow dumping 堆雪,雪垃圾
snow dune 雪丘
snow eater 吹在雪面上的暖风,雪面的雾,融雪风

snow effect 雪效应
snow erosion (nivation) 雪蚀
snow fall 降雪〔量〕
snow fence 雪栅栏
snow flea 冰跳蚤,冰蚤
snow flurry 小阵雪
snow forest climate 雪林气候
snow garland 雪花环
snow gauge 量雪器
snow grain 雪粒
snow grain size 雪粒尺寸
snow igloo 雪屋
snow injury 雪害
snow interception 降雪截留
snow leopards (*panthera uncia*) 雪豹
snow level 〔量〕雪尺
snow line 雪线
snow load 雪载荷
snow load pressure 雪压(力)
snow lotus (*saussurea involucrata*) 雪莲花
snow machine 雪车
snowman 雪人
snow map 积雪图
snow mass 积雪质量
snow mat 雪席
snow measuring plate 测雪板
snow melt 融雪
snow melt runoff 融雪径流
snow meltwater 融雪水
snow metamorphism 雪变质作用
snow metamorphosis 雪变质

snow mist 雪霭
snow model 积雪模式,积雪模型
Snow Model Intercomparison Project(SnowMIP) 雪模式比较计划
snow monitoring 积雪监测
snow moto 雪地摩托
snow niche (= nivation hollow) 雪蚀龛,雪蚀洼地
snow pack 积雪场
snow parameter 积雪参数
snow patch 雪斑
snow pellet (= tapioca snow) 霰(直径 2~5 mm),小雪球
snow permeability 雪渗透率
snow petrel (*pagodroma nivea*) 雪海燕
snow pigeon (*columba leuconota*) 雪鸽
snow pillow 雪水当量仪,雪枕
snow pit 雪坑
snow plow (= snow plough) 铲雪机
snow plume (= snow banner = snow smoke) 雪羽
snow pressure 雪压
snow profile 雪剖面
snow protection plantation 防雪林
snow quake (= snow tremor) 雪震
snow reflectivity 积雪反射率
snow reishi mushroom (*arenaria brevipetala*) 雪灵芝
snow removal 除雪
snow ringless amanita (*amanita nivalis*) 雪白鹅膏菌
snow ripple 雪纹
snow roughness length 积雪粗糙度
snow runway (飞机)雪跑道
snow sample 雪样
snow sample cutter 雪样铲
snow sampler 积雪采样器
snow saw 雪锯
snow sculpture 雪雕
snow season 雪季
snow shed 雪棚,雪挡
snow shield 防雪挡,防雪设备
snow shovel 雪锹
snow shower 阵雪
snow sky (= snow blink) 雪映白光,雪映天
snow slide 雪崩
snow sludge 湿雪泥
snow smoke (= snow banner = snow plume) 雪烟,雪羽,雪旗
snow spectrum 雪光谱
snow spot 雪斑
snow squall (= snowsquall) 雪飑
snow stability 积雪稳定性
snow stage 成雪阶段
snow stake 测雪花杆
snow static (电磁学)降雪静电
snow storm 雪暴
snow stratification 积雪成层作用
snow stratigraphy 雪层剖面
snow sublimation 积雪升华
snow surface temperature 雪面

温度
snow survey （路线）测雪
snow temperature 雪温
snow thermal insulation 雪隔热作用
snow thrower (＝snow blower) 扫雪机,螺旋桨式除雪机
snow trail 雪面小路
snow trapping capacity 雪捕捉力
snow tremor (＝snow quake) 雪震
snow tube 取雪管
snow type 雪型
snow tyre 雪地防滑轮胎
snow vehicle 雪地车
snow virga 雪幡
snow viscosity 雪黏度
snow water equivalent (SWE) 雪水当量
snow water sensor (SWS) 雪水当量仪
snow wetness 雪湿度
snow wreath 风积雪堆
snow zone 雪带
snowball 雪球
snowbelt 雪带
snowboarder 单板滑雪者
snowboat 船形雪橇
snow-catchment area 集雪区
snow-depth telemeter 雪深遥测计
snowdrift intensity 吹雪强度
snowfall intensity 降雪强度
snow-fed river 雪水补给河流
snowfield ski tour 雪地滑雪旅游

snowflake 雪花
snow-ice chemistry 雪冰化学
snow-line elevation 雪线高程
snowmaking 人工造雪
snowmelt erosion 融雪侵蚀
snowmelt excess 产流融雪水
snowmelt flood 融雪洪水
snowmelt forecast 融雪预报
snowmelt process 融雪过程
snowmelt runoff 融雪径流
snowmelt runoff forecast 融雪径流预报
snowmelt runoff model 融雪径流模型
snowmelt-fed stream 融雪补给河流
snowmelting rate 融雪速率
snowpack yield 积雪产流
snowpack (snow pack) 积雪场,雪堆
snowshed 防雪棚
snowline altitude 雪线高度
soaked snow (＝slush) 雪浆
social adaptability 社会适应力
social aspect 社会要素
sodium 钠
soft bed （冰川）软底
soft hail 软雹
soft windpack snow 软风板雪
soil auger 土壤钻,土钻
soil biogeochemistry 土壤生物地球化学
soil biomass 土壤生物量

soil characteristics 土壤特征
soil churning 冻卷作用
soil classification 土壤分类
soil climate 土壤气候
soil cohesion 土壤黏结力
soil colloid 土壤胶体
soil composition 土壤组成
soil density 土壤密度
soil dilation 土壤膨胀
soil ecology 土壤生态学
soil ecosyetem 土壤生态系统
soil erosion 土壤侵蚀
soil erosion damage 土壤侵蚀危害
soil erosion factor 土壤侵蚀因子
soil erosion index 土壤侵蚀指标
soil erosion intensity 土壤侵蚀强度
soil erosion modulu 土壤侵蚀模数
soil erosion monitoring 土壤侵蚀监测
soil evaporation 土壤蒸发
soil fauna 土壤动物区系
soil fertility 土壤肥力,土地肥沃度
soil freeze 土壤冻结
soil geography 土壤地理学
soil heat flux 土壤热通量
soil heating capacity 土壤热容量
soil horizon 土层
soil humidity 土壤湿度
soil humus 土壤腐殖质
soil hydroscopic water 土壤吸附水
soil loss amount 土壤流失量
soil mantle 土壤风化层
soil metabolism 土壤代谢
soil microbial biomass 土壤微生物量
soil microorganism (＝soil microbe) 土壤微生物
soil mineral 土壤矿物
soil moisture 土壤湿度,土壤水分
Soil Moisture and Ocean Salinity (SMOS) 土壤水分与海洋盐分卫星(欧空局)
soil moisture characteristics 土壤水分特征
soil moisture content 土壤含水量
soil moisture deficit 土壤缺水量
soil moisture gradient 土壤水分梯度
soil moisture profile 土壤湿度剖面
soil moisture regime 土壤水分状况
soil moisture stress 土壤水分胁迫
soil nutrient 土壤养分
soil organic matter 土壤有机质
soil organism (geobiont) 土壤生物
soil parent material 土壤母质
soil particle 土壤颗粒
soil permeability 土壤渗透性
soil physical property 土壤物理性质
soil plasticity 土壤可塑性
soil pore water 土壤孔隙水
soil porosity 土壤孔隙度
soil property 土壤特性

soil relative age   土壤相对年龄
soil respiration   土壤呼吸
soil salinity   土壤盐度
soil salt content   土壤含盐量
soil scour   土壤冲刷
soil seed bank   土壤种子库
soil structure (= soil constitution)
   土壤结构
soil surface thermometer   地表温度表
soil temperature   土壤温度
soil temperature regime   土壤温度状况
soil thermal exchange   土壤热交换
soil thermograph   土壤温度计
soil thermometer   土壤温度表
soil water   土壤水分
soil water balance   土壤水分平衡
soil wedge casts   土楔假型
soil-plant-atmosphere continuum (SPAC)   土壤－植物－大气连续体
soil-vegetation-atmosphere transfer (SVAT)   土壤－植被－大气传输
soil-water zone (= vadose zone)   土壤水层
solar constant   太阳常数
Solar Mesosphere Explorer (SME)   太阳－地球中层大气探险者卫星
solar radiation   太阳辐射
solar radiometer   太阳辐射表
solid precipitation   固态降水

solid state   固态
solidification   凝固
solidifying point   凝固点
solifluction   冻融泥流,泥流作用
solifluction apron   冻融泥流裙
solifluction fan   冻融泥流扇
solifluction feature   冻融泥流现象
solifluction flow   冻融泥流
solifluction lobe   冻融泥流舌
solifluction sheet   冻融泥流层
solifluction stripe   冻融泥流条带
solifluction terrace   融冻泥流阶地
soluble ions   可溶性离子
solute flux   溶质通量
Southern Oscillation (SO)   南方涛动
space photogrammetry   航天摄影测量
space photography   航天摄影
space remote sensing   航天遥感
space shuttle   航天飞机
space-based observation   空基观测
spacecraft   航天器
spatial resolution   空间分辨率
special report on emission scenarios (SRES)   排放情景特别报告
Special Sensor Microwave/Imager (SSM/I)   特殊传感器微波成像计
specific humidity   比湿
specific surface area (SSA)   比表面积
specific yield   给水度,单位出水量
speckle   (雷达图像)斑点

spectral albedo 谱反照率
spectral characteristic curve 光谱特性曲线
spectral resolution 光谱分辨率
spectrally intergrated flux extinction（SIFE） 谱综合通量衰减
spectrally intergrated snow albedo（SISA） 雪面谱综合反照率
spectrograph 谱仪
spectrometer 波谱测定仪
spectrophotometry 分光光度计
spectrum 谱
spectrum emissivity 谱发射率
specular reflection 镜面反射
speed skating 速滑（体育竞技项目）
sphagnic fibristels 泥炭鲜纤维有机寒冻土
spherical albedo 球面反照率
spillway 泄洪通道
spin up （模式）初始化运行
Spin-Scan Cloud Cover Camera（SSCCC） 自旋扫描云覆盖摄影机
spit 微量降水
spitze 冰镐前端
spodic horizon 灰化淀积层
spodic psammorthels 灰化淀积粗粒正常寒冻土
spodic psammoturbels 灰化淀积粗粒正常扰动寒冻土
spodic umbri-gelic cambosols 灰化暗瘠寒冻雏形土
spodosol 灰土
spongy ice 海绵冰
spontaneous condensation 瞬时凝结
spontaneous freezing 瞬时冻结
spontaneous sublimation 瞬时升华
sporadic permafrost 零星多年冻土
spore 孢子
sporopollen diagram 孢粉图式
sporopollen spectrum 孢粉谱
spring-fed stream 泉水补给河流
sprout 抽条，萌蘖，发芽，籽苗，萌芽，嫩芽，枝
sprout clump 丛生萌芽
sprouting phase 萌芽期
squeeze moraine 挤压冰碛
squirrel 松鼠
stable boundary layer（SBL） 稳定边界层
stable isotope 稳定同位素
stable stage-discharge relation 稳定水位流量关系
stadial （冰期内）冰段
stadial moraine（＝recessional moraine） 退缩碛垄，后退碛垄
stage gauging station 水位站
stage-discharge curve 水位流量曲线
stage-discharge relation 水位流量关系
stagnant glacier 停滞冰川
stagnant ice 停滞冰
stagnic umbri-gelic cambosols 滞水暗瘠寒冻雏形土
stake 花杆，测杆
stakeholder 利益有关方，利益有

关者
stalactite 钟乳石
stalagmite 石笋
stamukha 搁浅冰,落地冰
standard atmosphere 标准大气
standard atmosphere pressure 标准大气压
standard deviation 标准差
Standard Mean Ocean Water (SMOW) 标准平均大洋水（国际氢氧同位素标准）
standing biomass 现存生物量
state vector 状态向量
static chamber-gas chromatographic techniques (SCGCT) 静态箱—气相色谱观测系统
static cryosol 静态冷生土壤
static ice force 静冰力
static Poisson's ratio 静泊松比
statistical downscaling 统计降尺度
statistical indicator 统计指标
statistical survey 统计调查
statoscope 高差仪
steady infiltration 稳定渗透
steady state 稳定状态
steady turbulence 定常湍流
steam drill 蒸汽钻
steep ice ramp 陡峭冰坡面
Stefan's equation 斯特藩公式
Stefan-Boltzmann constant 斯特藩—玻尔兹曼常数
Stefan-Boltzmann law 斯特藩—玻尔兹曼定律
Stele Glacier 斯蒂尔冰川

stellar camera 恒星摄影机
stem stream 干流
steppe （半干旱气候下）草原
steppe shrub 草原灌丛
stereocamera (=stereometric camera) 立体摄影机
stereocomparator 立体坐标量测仪
stereographic projection 球面投影
stereometric camera (=stereocamera) 立体摄影机
stereophotogrammetry 立体摄影测量
stereoscope 立体镜
stereoscopic map 视觉立体地图,立体地图,立体图
stereoscopic model 立体模型
stereoscopic observation 立体观测
stereoscopic vision 立体视觉
stiffness 刚度,刚性
stimulate factor 激励因素
stochastic hydrology 随机水文学
stock-map 林相图
stomatal conductance 气孔导度
stomatal transpiration 气孔蒸腾
stone polygon soil 石质构造土,石质多边形土
stone rose 石玫瑰
stone stream 石河
stone-banked (solifluction) lobe 石质边缘的泥流舌
stony desert (=gravel desert) 砾漠
stony desertification 石漠化
stony earth circle 含碎石的土环（几何形土）

storm center 风暴中心
storm flood 暴雨洪水
storm-surge 风暴潮
stormwater model 雨洪模型
strain rate 应变率
stranded ice (= grounded ice) 冻透冰(河、湖、海岸冰),搁浅冰
stratified crown 分层型树冠
stratified drift 层状冰碛
stratified relief 层状地形
stratified sampling 分层抽样
stratigraphy 地层学
stratopause 平流层顶
stratosphere 平流层
Stratospheric Aerosol and Gas Experiment (SAGE) 平流层气溶胶和气体试验
stratovolcano 成层火山
streamflow 河流流量
streamline 流线
stress deviator 应力偏量
stress resistance 抗逆性
stress tensor field 应力张量场
stress tolerance 逆境耐性
stress-strain relationship 应力应变关系
striated boulder 擦痕巨砾
striated pebble 冰川条痕石
striation (冰川)条痕,(冰川)擦痕
striation stone 擦痕石
strontium isotope dating 锶同位素测年
structural terrace 构造阶地
structure of lava 熔岩构造

subalpine altoherbiprata 亚高山草群落
subalpine flora 亚高山植物区系
subalpine meadow 亚高山草甸
subalpine plant community 亚高山植物群落
subantarctic front 亚南极锋
subantarctic mode water 亚南极型水
subantarctic upper water 亚南极上层水
subantarctic zone 亚南极带
subarctic climate (= taiga climate) 亚北极地气候
subarctic current 副北冰洋流
subarctic forest 亚北极森林
subarctic intermediate water 亚北极中层水
subarctic region 亚北极区
subarctic water 亚北极水
subarctic zone 亚北极带
subatlantic period 亚大西洋期
subbasin 子流域
subboreal period 亚北方期
sub-continental glacier 亚大陆型冰川
subdominant species 亚优势种
subfrigid zone 亚寒带
subglacial 冰下的
subglacial ablation 冰下消融
subglacial cavity 冰下空洞
subglacial channel 冰下河道
subglacial debris 冰下岩屑
subglacial deposit 冰下沉积物

subglacial deposition 冰下沉积	subglacial water flux 冰下水流通量
subglacial drainage 冰下水系	subglacial water pressure 冰下水压
subglacial erosion 冰下侵蚀	
subglacial eruption 冰下喷发	subglacier lake 冰下湖
subglacial hydrologic system 冰下水文系统	sublacustrine 湖下的
	sublimation 升华
subglacial lake 冰下湖泊	sublimation ice 升华冰
subglacial landform 冰下地形	sublimation till 升华碛
subglacial melt 冰下消融	sublimatory metamorphism 升华变质
subglacial melt-out till 冰下融出碛	submarine pingo 海底冻胀丘
subglacial meltwater 冰下融水	submerged coast 下沉海岸
subglacial moraine 底碛	subpermafrost aquifer 多年冻土层下的含水层
subglacial mountains 冰下山脉	
subglacial outburst flood 冰下突发洪水	subpermafrost water 多年冻土层下水
subglacial permafrost 冰下多年冻土	subpixel 亚像元
subglacial plateau 冰下高原	subpolar glacier 亚极地冰川
subglacial process 冰下过程	subpolar gyre 亚极地环流
subglacial relief 冰下地形	subsea permafrost 海底多年冻土
subglacial ridge 冰下山脊	subsea talik 海底融区
subglacial runoff 冰下径流	subsidy 补贴
subglacial sediment 冰下沉积物	subsurface discharge 地下径流
subglacial sediment deformation 冰下沉积变形	subsurface exploration 地下勘察
	subsurface flow 壤中流
subglacial stream 冰下河流	subsurface irrigation 地下灌溉
subglacial talik 冰下融区	subsurface organic layer 次表层有机质层
subglacial till type 冰下冰碛类型	
subglacial topography 冰下地形	subsurface roughness 亚表面粗糙度
subglacial trough 冰下槽谷	
subglacial tunnel 冰下隧道	subsurface runoff 壤中流
subglacial valley 冰下谷地	subtropical (＝semitropical) 亚热带的,副热带的,半热带的
subglacial volcano 冰下火山	
subglacial water 冰下水	

subtropical high   副热带高压
subwater talik   水下融区
successional cropping   轮作,轮种
sulfate   硫酸盐
sulfate aerosols   硫酸盐气溶胶
sulfate reducing bacteria   硫酸盐还原菌
sulfite   亚硫酸盐
sulfur compound   硫化物
sulfur cycle   硫循环
sulfur dioxide（SO_2）   二氧化硫
sulfur hexafluoride（SF_6）   六氟化硫
sulfuric aquiturbels   硫化含水扰动寒冻土
sulfuric aquorthels   硫化物含水正常寒冻土
sulfuric horizon   含硫层
sulphate aerosol   硫酸盐气溶胶
sulphur oxides（SO_x）   硫氧化物
summer accumulation   夏季积累量
summer balance   夏平衡
summer monsoon   夏季风
sunlight   日照,日光
sunlight bleaching   （测年信号）阳光晒退
sunphotometer   太阳光度计
sunscald   灼伤
sunshine duration   日照时数
sunshine recorder   日照计
sunspot   太阳黑子
sun-synchronous orbit   太阳同步轨道
sun-synchronous satellite   太阳同步卫星
supercapillary interstice   超毛细管空隙
supercapillary percolation   超毛细管渗透
supercapillary porosity   超毛细管孔隙率
supercooled cloud   过冷云
supercooled vapor   过冷水汽
supercooled water droplet   过冷水滴
supercooling   过冷作用
superglacial deposition   冰面沉积作用
superglacial lake   冰面湖
superimposed ice   附加冰
superimposed ice zone   附加冰带
superpermafrost water   冻结层上水
superposed flow   叠加流
supersaturation   过饱和
supervised classification   监督分类
supraglacial meltwater   冰面融水
supraglacier lake   冰面湖
suprapermafrost water   多年冻土层上水
surf   海浪,涌浪
surface ablation   表面消融
surface adsorption   表面吸附
surface boundary condition   表面边界条件
surface boundary layer   近地边界层
surface coarse gravel   表面粗砾石
surface crack   表面裂纹

surface current 表层流
surface diffusion 表面扩散
surface displacement 表面位移
surface energy balance 地面能量平衡
surface energy budget 地面能量收支
surface erosion 面蚀
surface exposure dating 表面暴露测年
surface feature 表面特征
surface freezing index 地表冻结指数
surface geometry 表面形态
surface heating 表面加热
surface hoar 表面霜
surface inversion layer 近地逆温层
surface moraine 表碛
surface radiation balance 地表辐射平衡
surface resistance 表面阻力
surface roughness 表面粗糙度
surface runoff 地表径流
surface scattering 面散射
surface slope 表面坡度
surface strain 表面应变
surface streamline 表面流线
surface stress 表面应力
surface synoptic chart 地面天气图
surface synoptic station 地面天气站
surface temperature 表面温度
surface tensor 表面张力
surface thawing index 地表融化指数
surface thermometer 地面温度表
surface till 表碛物
surface undulation 表面起伏
surface water recharge 地表水补给
surface wind 地面风,海面风
surficial （地质)地表的,地面的
surge 跃动,涌
surging glacier 跃动冰川
survey line 测线
survey mark 测量标志
survey specifications 调查规范
survey technique 测量技术,测量方法
survival rate 成活率
suscitic cryogenic fabric 粒状冷生微结构
suspended matter 悬浮物
suspended particulate 悬浮颗粒物
suspended valley 悬谷
suspensoid 悬浮体
sustainability indicator 可持续性指标
sustainable development 可持续发展
suture zone （板块)缝合带
swamp 沼泽湿地
swamp facies 沼泽相
swath width （扫描)条带宽度
sweep time 扫描时间
swell mound 膨胀丘
swelling soil 膨胀土
symbiont 共生生物
symbiosis 共生
symbiotic nitrogen-fixing organism 共生固氮生物

symmetrical fold 对称褶皱
synchronism 同步性
Synchronous Meteorotogical Satellite (SMS) 同步气象卫星
synchronous satellite 同步卫星
synchronous teleconnection 同步遥相关
syngenetic ice 共生冰
syngenetic ice wedge 共生冰楔
syngenetic permafrost 共生多年冻土
synoptic 天气图的,天气尺度的
synoptic process 天气过程
synoptic system 天气系统
synthetic aperture radar (SAR) 合成孔径雷达
Synthetic Aperture Radar Interferometry (SAR Interferometry) 合成孔径雷达干涉测量
synthetic hydrogeological map 综合水文地质图
synthetic interferometer radar 合成干涉雷达
synthetic unit hydrograph 综合单位线(水文)
system ecology 系统生态学
system intergration 系统集成
systematic botany 植物系统分类学
systematic error 系统误差
systematic sample 系统样本
systematic sampling 系统抽样
systematic-random sampling 系统随机取样

T

tabetisol 融区(俄语)
table iceberg 桌状冰山
tabular ice 平顶冰
tabular iceberg (＝tabular berg) 平顶冰山
taele 冻土
tafoni 侵蚀孔,风化穴
taiga (＝boreal forest) 泰加林
taiga climate 泰加林气候,副极地气候
taiga snow 针叶林积雪
Tajik Nationality 塔吉克族
takyr 泥漠,龟裂土
takyric hapli-cryic aridosols 龟裂简育寒性干旱土
talik 融区
talus 坡积物
talus glacier 岩屑冰川
talus scree 倒石锥,岩屑堆
tangential adfreeze stress 切向冻结应力
tangential frost-heave force 切向冻胀力
tangut rhodiola (*rhodiola algida*)

唐古特红景天
tape ladder（=rope ladder） 小挂绳梯
tapioca snow（=snow pellet） 霰（直径2～5 mm），小雪球
tarn 封闭湖，小湖
Tatar Nationality 塔塔尔族
Taylor Dome （南极）泰勒冰穹
telescoped ice（=rafted ice） 叠加海冰
Television and Infra-Red Observation Satellite（TIROS） "泰罗斯"卫星（电视和红外观测卫星）
tamperate climate（=mesothermal climate） 温带气候
tamperate cold deciduous forest 温带冷落叶林
tamperate deciduous forest biome 温带落叶林生物群系
tamperate glacier 温冰川
tamperate ice 温冰
tamperature anomaly 温度异常，温度距平
tamperature contour（=isotherm） 等温线
tamperature dependence 温度依赖性
tamperature gradient 温度梯度
tamperature inversion 逆温
tamperature profile 温度剖面
tamperature-salinity diagram 温盐曲线
tamporal resolution 时间分辨率

tamporal scale 时间尺度
tamporary settlement 临时居住点
tensile fracture 拉伸断裂，张断面
tensile strain 拉伸应变
tensile strength 抗拉强度
tensile stress 拉应力，张应力
tensional crack 张裂隙
tensor field 张量场
tented ice 拱形冰堆
tephra 火山玻璃
tephrochronology 火山灰年代学
tephrostratigraphy 火山灰地层学
terminal ice 末端冰
terminal moraine 终碛垄
terminal retreat 末端后退
terminus （冰川）末端
terrace 阶地，梯田
terrain geocoding 地形编码
terrestrial deposit 陆相沉积
terrestrial ecosystem 陆地生态系统
terrestrial in situ cosmogenic nuclide（TCN） 陆地上就地宇宙成因核素
terrestrial radiation 地面辐射
terrestrial stereo photogrammetry 地面立体摄影
terric fibric organic cryosol 矿底纤维有机寒冻土
terric fibri-permagelic histosols 矿底纤维多年冻结有机土
terric fibristels 矿底纤维有机寒冻土
terric hemi-permagelic histosols

矿底半腐多年冻结有机土
terric hemistels 矿底半腐有机寒冻土
terric humic organic cryosol 矿底腐殖有机寒冻土
terric mesic organic cryosol 矿底中湿有机寒冻土
terric sapristels 矿底高腐有机寒冻土
terrigenous clastic rocks 陆源碎屑
terrigenous lake 陆源湖,陆成湖
terrigenous sediment 陆源沉积物
Tertiary 第三纪
Tethys Sea 特提斯海
thaw 融化
thaw basin 融沉盆地
thaw collapse 融陷
thaw compressibility 融化压缩性
thaw compressibility coefficient 融化压缩系数
thaw concave 融蚀凹地
thaw consolidation 融化固结
thaw consolidation ratio 融化固结率
thaw depression 热融洼地
thaw depth 融化深度
thaw hole 热融孔
thaw lake 热融湖
thaw lake talik 热融湖融区
thaw penetration 融渗
thaw period 融化期,解冻期
thaw puddle 积水热融洼地
thaw settlement 融化下沉
thaw settlement sensitivity 融沉敏感性
thaw sink 热融坑
thaw sinkhole 热融洞穴
thaw slumping 热融滑塌
thaw stability 融化稳定性
thaw strain 融化应变
thaw subsidence 融化下沉
thaw unconformity 融化不整合
thaw weakening (＝thaw softening) (冻土强度)热融弱化
thawed ground (＝thawed soil) 融土
thawed soil (＝thawed ground) 融土
thawing erosion 融冻侵蚀
thawing front 融化锋面
thawing season 解冻季节
thawing sedimentation 热融沉积
thawing soil 正融土
thawing subsidence character 融沉特征
thawing time 融化时间
thaw-sensitive permafrost 融化敏感性多年冻土
thaw-settlement coefficient 融化下沉系数
thaw-stable frozen soil 融化稳定性冻土
thaw-unstable frozen soil 融化不稳定性冻土
thematic map 专题地图
Thematic Mapper (TM) 专题制图仪(美国)
theodolite (＝transit) 经纬仪
thermal capacity (＝heat capacity) 热容量
thermal circulation 热环流

thermal conductivity 热传导率
thermal contraction cracking 温缩裂缝
thermal convection 热对流
thermal cracking 热开裂
thermal diffusion 热扩散
thermal diffusion coefficient (= thermal diffusivity) 热扩散系数
thermal diffusivity (= thermal diffusion coefficient) 热扩散系数
thermal dissipation 热耗散
thermal disturbance 热扰动
thermal equilibrium 热量平衡
thermal equivalent 热当量
thermal erosion 热力侵蚀
thermal expansion 热膨胀
thermal filter 热过滤器
thermal graph 热图像
thermal infrared 热红外
thermal infrared image 热红外成像
thermal infrared radiation 热红外辐射
thermal inversion layer 逆温层
thermal offset 温度位移
thermal pollution 热污染
thermal potential 热位势
thermal probe 热探头
thermal property 热力性质
thermal radiation 热辐射
thermal regime 热状态
thermal resistance 热阻抗
thermal semi-conductor 热半导体
thermal stability 热稳定性
thermal stress 热应力
thermal structure 热结构
thermal suffosion 热潜蚀
thermal talik 热融区
thermal undercutting 热蚀
thermal-contraction-crack ice 热缩裂冰
thermistor 热敏电阻〔温度计〕
thermocirques 热融蚀斗
thermocline 温跃层
thermocline layer 斜温层,变温层
thermocouple 热电偶〔温度计〕
thermocouple dew point hygrometer 热电偶露点湿度表
thermodynamics 热力学
thermo-erosional cirque 热融斗
thermo-erosional niche 热融壁龛
thermography 自计温度,测温法
thermohaline circulation 温盐环流,热盐循环
thermohygrograph 温湿计
thermokarst 热喀斯特
thermokarst depression 热融洼地
thermokarst lake 热喀斯特湖
thermokarst landform 热喀斯特地形
thermokarst mound 热喀斯特丘
thermokarst pond 热喀斯特池塘
thermokarst subsidence 热喀斯特沉陷
thermokarst terrain 热喀斯特台地
thermokarst topography 热喀斯特地形

Thermoluminescence (TL) dating 热释光测年
thermometer 温度计
thermoplanation 热融夷平作用
thermoplanation terrace 热融夷平阶地
thick ground ice 厚层地下冰
thick lake ice 厚湖冰(厚度 30~70 cm)
thicket 灌木丛
thin film 薄膜
thin lake ice 薄湖冰(厚度 5~15 cm)
threshold 阈值
threshold wind velocity 启动风速,临界风速
through glacier 分叉冰川
thufur 冰土丘,冻融草丘
thundersnow (=thunder snowstorm) 雪雷暴
tibetan antelope (*pantholops hodgsoni*) 藏羚羊
Tibetan Buddhist 藏传佛教
tibetan gazelle (*procapra picticaudata*) 藏原羚
tibetan high 青藏高压
Tibetan Nationality 藏族
tibetan tourism 西藏旅游
tidal crack 潮汐破裂(海冰)
tidal glacier (=tidewater glacier) 入海冰川
tidewater glacier (=tidal glacier) 入海冰川
till 冰碛
till base 冰碛底部
till billow 冰碛脊
till fabric 冰碛组构
till layer 冰碛层
till plain 冰碛平原
till plain landform 冰碛平原地形
till sheet 冰碛层
till wall 冰碛墙,冰碛崖
till-cored esker 冰碛核蛇丘
tillite 冰碛岩
timber line (=timberline) 森林界线,林线
Time Domain Reflectometry (TDR) 时域反射仪
time resolution 时间分辨率
TIROS Operational Satellite (TOS) "泰罗斯"业务卫星
toboggan 平底雪橇
tongue-shaped rock glacier 舌状石冰川
top of atmosphere (TOA) 大气层顶
topographic correction 地形校正
topographic map 地形图
topographic survey 地形测量
topography 地形学
toponomastics (=toponymy) 地名学
total dew point 总露点
total ice force 总冰力
total net budget 总净收支
total nitrogen 全氮
total ozone 臭氧总量
Total Ozone Mapping Spectrometer (TOMS) 臭氧全量图分光计
total primary productivity 总初级

生产力
total runoff 总径流量
Total Suspended Particles (TSP) 总悬浮颗粒物
tourism 旅游业,旅游
trace 痕量
trace element 痕量元素
trace gases 痕量气体
tracer 示踪物,示踪剂
Track and Data Ralay Satellite System (TDRSS) 跟踪和数据中继卫星系统
Track-Cross Track-Normal (reference system)(TCN) 雷达坐标系统(航向－交叉向－法向)
Trans-Antarctic Mountains 南极横断山脉
transboundary impact 跨境影响
transboundary river 跨境河流
transboundary waters 跨境水域
transgression 海侵,海进
transgressive-regressive sequence 海侵海退层序
transitional zone 过渡区,过渡带
transitory permafrost 过渡型多年冻土
transparency 透明度,透明性
transpiration 蒸腾〔作用〕
transpiration coefficient 蒸腾系数
transpiration efficiency 蒸腾效率
transpiration flux 蒸腾通量
transpiration rate 蒸腾速率
transpiration ratio 蒸腾比
transport capacity 搬运能力,运输能力
transport corridor 交通走廊
transport cost 交通成本
transport pathway 传输路径
transported deposit 搬运沉积,运积层
travel potential market 旅游潜在市场
travois 雪橇(印第安语)
tree age 树龄
tree line 树线,林线
Tree Preservation Oder (TPO) 护林法案
tree ring 树轮
Triangular Irregular Network (TIN) 不规则三角形网络
tributary 支流
tributary glacier 支冰川
trigonometric leveling 三角高程测量
tritium 氚
tropopause 对流层顶
troposphere 对流层
trough 槽,槽谷
trough valley (= U-shaped valley) 槽谷,U型谷
true color image 真彩色影像
true meridian 真子午线
trunic halpi-gelic cambosols 表蚀简育寒冻雏形土
trunk glacier 主干冰川
tsamba 糌粑
Tu Nationality 土族
tuekahoe (*wolfiporia cocos*) 茯苓
tundra 苔原,冻原

tundra belt　苔原带,冻原带
tundra climate　苔原气候
tundra desert　苔原荒漠
tundra ecosystem　苔原生态系统
tundra fauna　苔原动物群
tundra flora　苔原植物群
tundra landform　苔原地貌
tundra landscape　苔原景观
tundra polygon　苔原多边形
tundra snow　苔原积雪
tundra soil　苔原土
tundra vegetation　苔原植被
tundra wetland　苔原湿地
tundra zone　苔原带
turbels　扰动寒冻土亚纲
turbic cryosol　扰动寒冻土
turbosphere　湍流层
turbulence　湍流
turbulence condensation level（TCL）　湍流凝结高度
turbulent diffusion coefficient　湍流扩散系数
turbulent flux　湍流通量
turbulent kinetic energy　湍流动能
turbulent mixing　湍流混合
turbulent transfer coefficient　湍流传输系数
turbulent viscous coefficient　湍流黏性系数
turbulivity　湍流度,湍流系数
turf hummock　草丘
turf-banked lobe（terrace）　围草皮舌（阶地）
turret ice　侧立冰,塔冰

twin crystal　双晶〔体〕
twinned glacier　双支冰川,孪生冰川
two-color laser ranger　双色激光测距仪
two-layer permafrost　双层多年冻土
two-side freezing　双向冻结
Tyndall flowers（tyndall figure）　廷德尔冰花,冰面上六面形水坑
typic anhyorthels　普通无水正常寒冻土
typic anhyturbels　普通脱水扰动寒冻土
typic aqui-gelic cambosols　普通潮湿寒冻雏形土
typic aquiturbels　普通含水扰动寒冻土
typic aquorthels　普通含水正常寒冻土
typic argi-cryic aridosols　普通黏化寒性干旱土
typic argiorthels　普通黏化正常寒冻土
typic calci-cryic aridosols　普通钙积寒性干旱土
typic cryi-ustic isohumosols　普通寒性干润均腐土
typic fibri-permagelic histosols　普通纤维多年冻结有机土
typic fibristels　普通纤维有机寒冻土
typic foli-permagelic histosols　普通落叶多年冻结有机土

typic folistels 普通落叶有机寒冻土
typic geli-alluuvic primosols 普通寒冻冲积新成土
typic geli-cryic andosols 普通寒冻寒性火山灰土
typic geli-orthic primosols 普通寒冻正常新成土
typic geli-sandic primosols 普通寒冻砂质新成土
typic glacistels 普通厚冰层有机寒冻土
typic gypsi-cryic aridosols 普通石膏寒性干旱土
typic halpi-gelic cambosols 普通简育寒冻雏形土
typic hapli-cryic andosols 普通简育寒性火山灰土
typic hapli-cryic aridosols 普通简育寒性干旱土
typic hapli-permagelic gleyosols 普通简育多年冻结纤维土
typic haplorthels 普通弱育正常寒冻土
typic haploturbels 普通弱育扰动寒冻土
typic hemi-permagelic histosols 普通半腐多年冻结有机土
typic hemistels 普通半腐有机寒冻土
typic histi-permagelic gleyosols 普通有机多年冻结纤维土
typic historthels 普通有机正常寒冻土
typic histoturbels 普通有机扰动寒冻土
typic matti-gelic cambosols 普通草毡寒冻雏形土
typic molli-gelic cambosols 普通暗沃寒冻雏形土
typic molliturbels 普通暗沃扰动寒冻土
typic mollorthels 普通暗沃正常寒冻土
typic permi-gelic cambosols 普通多年冻结寒冻雏形土
typic psammorthels 普通粗粒正常寒冻土
typic psammoturbels 普通粗粒正常扰动寒冻土
typic sapristels 普通高腐有机寒冻土
typic umbri-gelic cambosols 普通暗瘠寒冻雏形土
typic umbriturbels 普通暗瘠扰动寒冻土
typic ummrorthels 普通暗瘠正常寒冻土
typical alpine meadow 典型高寒草甸

U

ultra high frequency（UHF） 特高频
ultraviolet ray 紫外线
umbrella effect 阳伞效应
umbric epipedon 暗瘠表层
umbriturbels 暗瘠扰动寒冻土
umbrothels 暗瘠正常寒冻土
unbroken ice 未破碎冰
unchecked development 盲目发展
under color addition 底色增益
under color removal 底色去除
undergrazing 轻度放牧
under-ice water layer 冰下水层
underlayer 下层,垫层,下伏层
underlying surface 下垫面
undermelting （浮冰）底部融化
unfreezing port 不冻港
unfrozen lake 未冻湖
unfrozen soil 未冻土
unfrozen water 未冻水
unfrozen water content 未冻水含量
unfrozen zone 未冻结带
uniaxial 单轴的
uniaxial compression 单轴压缩
uniaxial extension 单轴拉伸
United Nations Conference on Environment and Development（UNCED） 联合国环境与发展大会
United Nations Conference on the Human Environment（UNCHE） 联合国人类环境会议
United Nations Disaster Relief Office（UNDRO） 联合国救灾处
United Nations Framework Convention on Climate Change（UNFCCC） 联合国气候变化框架公约
Universal Polar Stereographic（UPS） Projection 通用极球面投影
Universal Transverse Mercator（UTM） Projection 通用横墨卡托投影
unsaturated air 未饱和空气
unsaturated downdraft 未饱和下沉气流
unsaturated flow 非饱和流
unstable air mass 不稳定气团
unstable condition 不稳定条件,不稳定状态
unstable equilibrium point 不稳定平衡点
unstable snow cover 不稳定积雪
unstable stratification 不稳定层结
U-Pb dating 铀铅法测年
upland 高地
upland swamp 高地沼泽
upper air circulation 高空环流
upper air current 高空气流
Upper Atmosphere Research Satellite（UARS） 高层大气研究卫星
Upper Continental Crust（UCC） 上地壳

upper forest limit 森林上限
upper-air sounding 高空探测
upper-level jet stream 高空急流
upscaling 升尺度
upslope wind 上坡风
upward flow 上升气流
upward freezing 向上冻结
upward motion 向上运动
upwelling 上升流(海洋)
upwind difference 迎风差分
Ural blocking high 乌拉尔山阻塞高压
Uranium-lead dating 铀铅年龄测定
Uranium-series dating 铀系测年
Uranium-Thorium dating 铀/钍测年
urban ecology 城市生态学
urban heat island 城市热岛
uric acid 尿酸
US Geological Survey（USGS） 美国地质调查局
U-shaped valley U形谷
Uygur Nationality 维吾尔族
Uzbek Nationality 乌兹别克族

V

vadose zone（＝soil water zone） 土壤渗流层
valley glacier 山谷冰川
valley-wall rock glacier 谷壁型石冰川
vapor phase 汽相
vaporation（＝vaporization） 汽化
vaporization heat 汽化热
vapor（＝vapour） 水汽
vapour（＝vapor） 水汽
varved clay 纹泥,季候泥
varved sediment （冰川）纹泥沉积
vector data 矢量数据
vegetation classification 植被分类
vegetation cover 植被覆盖
vegetation index 植被指数
vegetation regionalization 植被区划
vegetation survey 植被调查
vegetation variation 植被变化
vein ice 脉状冰
Venus Orbiting Imaging Radar（VOIR） 金星轨道成像雷达
vertic molliturbels 变性暗沃扰动寒冻土
vertic mollorthels 变性暗沃正常寒冻土
vertic umbriturbels 变性暗瘠扰动寒冻土
vertic umbrorthels 变性暗瘠正常寒冻土
vertical climatic zonality 垂直气候地带性

vertical control survey 高程控制测量
vertical ice force 垂向冰力
vertical polarization 垂直极化
vertical resolution 垂直分辨率
vertical zonality 垂直地带性
very close ice 极密集冰
Very High Resolution Radiometer (VHRR) 甚高分辨率辐射仪
Very Long Baseline Interferometry (VLBI) 甚长基线干涉测量
Vesuvius Volcano 维苏威火山
violent break-up river 武开河
virgin area (= primeval area) 原始〔植被〕区
virgin forest (= virgin wood-land) 原始林
viscidity 黏滞性
viscoplastic fluid 黏塑性流体
viscosity 黏滞度
viscosity force 黏性力
viscous flow 黏滞流,黏性流〔动〕
viscous relation 黏性关系
viscous shear 黏性剪力
visibility 能见度
Visible and Infra Red Radiometer (VIRR) （风云三号卫星的）可见光红外扫描辐射计
visible ice 可见冰
visible light 可见光
Visible/Infrared Spin Scan Radiometer (VISSR) 可见光/红外自旋扫描辐射仪
visible-near infrared (VIR) image 可见光近红外影像
visual observation 目测
visual representation 可视化表达
visual zenith telescope 目视天顶仪
vitrandic aquorthels 玻璃火山灰含水正常寒冻土
vitrandic molliturbels 玻璃火山灰暗沃扰动寒冻土
vitrandic mollorthels 玻璃火山灰暗沃正常寒冻土
vitrandic umbriturbels 玻璃火山灰暗瘠扰动寒冻土
vitrandic ummrorthels 玻璃火山灰暗瘠正常寒冻土
vitric hapli-cryic andosols 玻璃简育寒性火山灰土
void 空隙,空穴
volatile organic compounds (VOCs) 挥发性有机化合物
volcanic aerosol 火山气溶胶
volcanic ash 火山灰
volcanic ash layer 火山灰层
volcanic cinder 火山渣
volcanic ejecta 火山抛出物
volcanic eruption 火山喷发
volcanic fragmental rock 火山碎屑岩
volcanic gas 火山喷发气体
volumetric heat capacity 体积热容量
volumetric ice content 体积含冰量

volumetric water content 体积含水量
vorthern twinflower (*linnaea borealis*) 北极花
Vostock ice core 东方站冰芯(南极)
V-shaped valley V形谷
vulnerability 脆弱性

W

wady (=wadi) 河谷(季节河流)
wall-sided glacier 陡壁冰川
walrus (*arapaima gigas*) 海象
warm permafrost 高温多年冻土
warning system 预警系统
wash plain 淤积平原
wash-board moraine 冰水冰碛垄,波纹状冰碛
washboard road surface 搓板路面
washed drift 层状冰碛
washout rate 清洗率
wastage 损耗
water affinity 亲水性
water balance 水平衡,水量平衡
water consumption 耗水量
water contamination 水污染
water cycle 水循环
water equivalent 水当量
water erosion 水蚀
water flux 水流量
water holding capacity 持水能力
water ice 水冰
water potential 水势
water potential difference 水势差
water potential gradient 水势梯度
water pressure 水压
water protection zone 水资源保护区
water quality 水质
water resources 水资源
water sample 水样
water storage 蓄水
water table 水位
water tightness 密封性
water use efficiency (WUE) 水分利用效率
water vapor flux 水汽通量
water vapor pressure 水汽压
water vapor (=water vapour) 水汽
water vapour (=water vapor) 水汽
water yield property 产流特性
water-holding capacity 持水度
waterproof stratum 不透水层
watershed development 流域开发
watershed divide 分水岭
watershed ecosystem 流域生态系统
watershed scale (=catchment scale)

　　　　小流域尺度
water-soaked till　浸水冰碛
water-soluble ion　可溶离子
wave absorption　波吸收
wave amplitude　波幅
wave attenuation　波衰减
wave band　波段
wave base　波蚀基面
wave diffraction　波衍射, 波绕射
wave field　波场
wave filter　滤波器
wave frequency　波频率
wave interference　波干涉
wave number　波数
wavelet analysis　小波分析
wavelet transform　小波变换
weather station　天气站
weatherability　风化能力
weathered ridge　风化脊
weathering　风化作用
weathering crust　风化壳
weathering intensity　风化强度
weathering profile dating　风化剖面测年
weathering residue　风化残余物
Weertman-type sliding　威尔特曼型滑动
weighing snow-gauge　雪秤
weighted mean　加权平均数
well-bonded permafrost　胶结良好的多年冻土
West Antarctic Ice Sheet (WAIS)　西南极冰盖
wet deposition　湿沉降
wet snow　湿雪
wet snow line　湿雪线
wet snow zone　湿雪带
whaleback　鲸背石
white dew　冻露
white frost　白霜
whiteout　白化天气
Wide-angel infrared Dual-mode Line/area Array Scanner (WiDAS)　机载红外广角双模式成像仪
wiggle trace　波形曲线
wild snow　蓬松新雪
wilson's storm-petrel (*oceanites oceanicus*)　巨海燕
wind abrasion　风蚀
wind chill　风寒
wind crust　风壳, 风板
wind damage　风害
wind drag force　风拖曳力
wind drift sand flow　风沙流
wind erosion　风蚀
wind erosion landform　风蚀地貌
wind packed snow　风板
wind ripple　风雪纹
wind slab　风板
winter fast ice　冬季固定冰
winter ice　冬季冰
winter monsoon　冬季风
winter resistance　抗寒性, 耐寒性
winterization　防冻措施, 准备过冬
winter-talus ridge　冬季岩屑堆脊
wire icing　电线积冰

wireless sensor network（WSN） 无线传感器网络
Wisconsin Glaciation （北美）威斯康星冰期
wood sledge 木雪橇
World Geodetic System（WGS） 世界测绘坐标系
World Meteorological Day 世界气象日
World Meteorological Organization（WMO） 世界气象组织
World Water Day 世界水日
Würm Glaciation 玉木冰期

X

xerophilous（= xerophytic） 适旱的
xerophorbium 冻原
xerothermic period 干热期,干温期
Xibe Nationality 锡伯族
Xixiabangma Glaciation 希夏邦马冰期
X-ray photogrammetry X 射线摄影测量

Y

yak（*poephagus mutus*） （西藏）牦牛
Yi Nationality 彝族
young coastal ice 初岸冰
young ice 初冰
Younger Dryas（YD） 新仙女木期
Yugur Nationality 裕固族

Z

zero curtain 零点幕
zero emission 零排放

附录

附录2 国内外著名冰川

序号	英文名	中文名	所在地
1	Abramov Glacier	阿布拉莫夫冰川	【吉尔吉斯斯坦】
2	Academy Glacier	学院冰川	【格陵兰】
3	Academy Glacier	学院冰川	【南极洲】
4	Acosta Glacier	阿科斯塔冰川	【南极洲】
5	Alerce Glacier	阿勒斯冰川	【阿根廷】
6	Aletsch Glacier	阿莱奇冰川	【瑞士】
7	Alfotbreen Glacier	阿尔佛特伯林冰川	【挪威】
8	Allalin Glacier	阿拉林冰川	【瑞士】
9	Amalia Glacier	阿马利娅冰川	【智利】
10	Amery Ice Shelf	埃默里冰架	【南极洲】
11	Argentiere Glacier	阿根廷艾尔冰川	【法国】
12	Arrow Glacier	箭头冰川	【坦桑尼亚】
13	Artesonraju Glacier	阿特桑拉吉冰川	【秘鲁】
14	Arwa Glacier	阿娃冰川	【印度】
15	Athabasca Glacier	阿萨巴斯卡冰川	【加拿大】
16	Austdalsbreeen Glacier	奥斯达而伯林冰川	【挪威】
17	Austre Broeggerbreen Glacier	奥斯塔博拉格伯林冰川	【挪威】
18	AX010 Glacier	AX010 冰川	【尼泊尔】
19	Azha Glacier	阿扎冰川	【中国】
20	Baby Glacier	巴比冰川	【加拿大】
21	Bahia del Diablo Glacier	巴伊亚代阿布洛冰川	【南极洲】
22	Baishuihe Glacier No.1	白水河1号冰川	【中国】
23	Balmaceda Glacier	巴尔马塞达冰川	【智利】
24	Baltoro Glacier	巴尔托洛冰川	【巴基斯坦】
25	Bara Shigri Glacier	芭拉什格利冰川	【印度】

续表

序号	英文名	中文名	所在地
26	Barranco Glacier	巴兰科冰川	【坦桑尼亚】
27	Basodino Glacier	巴索迪诺冰川	【瑞士】
28	Batura Glacier	巴托拉冰川	【巴基斯坦】
29	Bearman Glacier	贝尔曼冰川	【南极洲】
30	Bellisime Glacier	贝利塞姆冰川	【南极洲】
31	Belukha Glacier	别卢哈冰川	【俄罗斯】
32	Bench Glacier	本奇冰川	【美国】
33	Bering Glacier	白令冰川	【美国】
34	Bernardo Glacier	伯纳德冰川	【智利】
35	Bhadal Glacier	巴达尔冰川	【印度】
36	Biafo Glacier	比亚夫冰川	【巴基斯坦】
37	Bishop Glacier	毕晓普冰川	【加拿大】
38	Blomstolskar Glacier	布拉姆斯托斯卡亚冰川	【挪威】
39	Blue Glacier	布卢冰川（蓝冰川）	【美国】
40	Bondhusbreen Glacier	伯德斯布林冰川	【挪威】
41	Breidalblikkbrea Glacier	布雷达尔不利卡布里冰川	【挪威】
42	Breidamjökull Glacier	布雷达姆马加库尔冰川	【冰岛】
43	Brenva Glacier	布朗瓦冰川	【意大利】
44	Brewster Glacier	布鲁斯特冰川	【新西兰】
45	Bruarjökull Glacier	布拉亚马加库尔冰川	【冰岛】
46	Brüggen Glacier	布鲁根冰川	【智利】
47	Brunegg Glacier	布朗内格冰川	【瑞士】
48	Bulbur Glacier	布尔波冰川	【南极洲】
49	Burroughs Glacier	波勒斯冰川	【美国】
50	Calderone Glacier	考尔德伦冰川	【意大利】
51	Careser Glacier	卡雷萨冰川	【意大利】
52	Carstensz Glacier	卡尔赞兹冰川	【印度尼西亚】

续表

序号	英文名	中文名	所在地
53	Casa Pangue Glacier	卡萨庞格冰川	【智利】
54	Castao Overo Glacier	卡斯塔奥欧瓦罗冰川	【阿根廷】
55	Cesar Glacier	恺撒冰川	【肯尼亚】
56	Chacaltaya Glacier	查卡塔亚冰川	【玻利维亚】
57	Chandra Glacier	钱德拉冰川	【印度】
58	Chandra Nahan Glacier	钱德拉纳哈冰川	【印度】
59	Charquini sur Glacier	查尔奎尼瑟尔冰川	【玻利维亚】
60	Chavez Glacier	查韦斯冰川	【南极洲】
61	Chernov Glacier	切尔诺夫冰川	【俄罗斯】
62	Chhota Shigri Glacier	查塔什格力冰川	【印度】
63	Chong Kumdan Glacier	崇昆丹冰川	【印度】
64	Chongce Glacier	崇侧冰川	【中国】
65	Ciardoney Glacier	西亚德尼冰川	【意大利】
66	Coleman Glacier	科尔曼冰川	【美国】
67	Colonia Glacier	科洛尼亚冰川	【智利】
68	Columbia (2057) Glacier	哥伦比亚2057冰川	【美国】
69	Columbia Glacier	哥伦比亚冰川	【格陵兰】
70	Cook Ice Shelf	库克冰架	【南极洲】
71	Cooke Glacier	库克冰川	【南极洲】
72	Cosgrove Ice Shelf	卡斯格拉夫冰架	【南极洲】
73	Cox Glacier	考克斯冰川	【南极洲】
74	Craft Glacier	克拉夫特冰川	【南极洲】
75	Credner Glacier	克勒脱纳冰川	【坦桑尼亚】
76	Crosson Ice Shelf	克罗松冰架	【南极洲】
77	Daniels Glacier	丹尼尔斯冰川	【美国】
78	Darwin Glacier	达尔文冰川	【南极洲】
79	Dasuopu Glacier	达索普冰川	【中国】

续表

序号	英文名	中文名	所在地
80	Daugaard-Jensen Glacier	杜加尔德金森冰川	【格陵兰】
81	Deadmond Glacier	戴得蒙德冰川	【南极洲】
82	Decken Glacier	得肯冰川	【坦桑尼亚】
83	Devon Ice Cap	德文冰帽	【加拿大】
84	Diamond Glacier	钻石冰川	【坦桑尼亚】
85	Djankuat Glacier	德加库亚特冰川	【俄罗斯】
86	Dokriani Glacier	多克利亚尼冰川	【印度】
87	Dongkemadi Glacier	冬克玛底冰川	【中国】
88	Drygalski Glacier	德里加尔斯基冰川	【坦桑尼亚】
89	Dunde Glacier	敦德冰川	【中国】
90	Dyngjujökull Glacier	丁吉加库尔冰川	【冰岛】
91	Easton Glacier	伊斯顿冰川	【美国】
92	Echaurren Norte Glacier	依查伦诺特冰川	【智利】
93	Elena Glacier	埃琳娜冰川	【肯尼亚】
94	Elena Glacier	埃琳娜冰川	【乌干达】
95	Eliot Glacier	埃利奥特冰川	【美国】
96	Elisebreen Glacier	埃里斯伯林冰川	【挪威】
97	Emmons Glacier	埃蒙斯冰川	【美国】
98	Engabreen Glacier	安佳伯林冰川	【挪威】
99	Eugenie Glacier	尤金妮娅冰川	【加拿大】
100	Exum Glacier	埃克萨姆冰川	【南极洲】
101	Eyabakkajökull Glacier	依亚巴卡加库尔冰川	【冰岛】
102	Fedchenko Glacier	费德琴科冰川	【塔吉克斯坦】
103	Ferrar Glacier	费拉尔冰川	【南极洲】
104	Findelen Glacier	芬得兰冰川	【瑞士】
105	Foley Glacier	福利冰川	【南极洲】
106	Fontana Bianca Glacier	丰塔纳比安卡冰川	【意大利】

续表

序号	英文名	中文名	所在地
107	Forel Glacier	羊皮纸冰川	【肯尼亚】
108	Forni Glacier	福尼冰川	【意大利】
109	Foss Glacier	福斯冰川	【美国】
110	Fox Glacier	福克斯冰川	【新西兰】
111	Frankenfield Glacier	弗兰克菲尔德冰川	【南极洲】
112	Franz Josef Glacier	弗朗茨约瑟夫冰川	【新西兰】
113	Frébouge Glacier	弗雷伯格冰川	【意大利】
114	Freya Glacier	芙蕾雅冰川	【格陵兰】
115	Frias Glacier	弗里亚斯冰川	【阿根廷】
116	Furtwangler Glacier	富特文格勒冰川	【坦桑尼亚】
117	Gades Glacier	高狄斯冰川	【格陵兰】
118	Gangotri Glacier	根戈德里冰川	【印度】
119	Garabashi Glacier	加拉巴什冰川	【俄罗斯】
120	Gebroulaz Glacier	吉布娄拉斯冰川	【法国】
121	Gictro Glacier	吉克特罗冰川	【瑞士】
122	Gilkey Glacier	吉尔基冰川	【美国】
123	Glacier des Bossons	波松斯冰川	【法国】
124	Glacier No. 1 in headwater of Urümqi River	乌鲁木齐河源1号冰川	【中国】
125	Glacier No. 12 in Laohugou Valley	老虎沟12号冰川	【中国】
126	Glacier No. 5 in Yanglong River	羊龙河5号冰川	【中国】
127	Godwin-Austen Glacier	戈德温—奥斯汀冰川	【巴基斯坦】
128	Goff Glacier	戈夫冰川	【南极洲】
129	Goldbergkees Glacier	戈柏契斯冰川	【奥地利】
130	Goodell Glacier	古德尔冰川	【南极洲】

续表

序号	英文名	中文名	所在地
131	Gopher Glacier	地鼠冰川	【南极洲】
132	Gora Glacier	戈拉冰川	【印度】
133	Graafjellsbreen Glacier	格拉芙吉尔斯伯林冰川	【挪威】
134	Graasubreen Glacier	格拉苏伯林冰川	【挪威】
135	Grand Etret Glacier	格兰特爱特里特冰川	【意大利】
136	Gregory Glacier	格雷戈里冰川	【肯尼亚】
137	Gries Glacier	格里斯冰川	【瑞士】
138	Grinnell Glacier	格林内尔冰川	【美国】
139	Guliya Glacier	古里雅冰川	【中国】
140	Gulkana Glacier	库尔卡纳冰川	【美国】
141	Guxiang Glacier	古乡冰川	【中国】
142	Hailuogou Glacier	海螺沟冰川	【中国】
143	Hale Glacier	黑尔冰川	【南极洲】
144	Hamta Glacier	哈姆塔冰川	【印度】
145	Hans Glacier	汉斯冰川	【挪威】
146	Hansbreen Glacier	汉斯布林冰川	【挪威】
147	Harald Moltke Glacier	哈拉尔德莫尔特克冰川	【格陵兰】
148	Hardangerjoekulen Glacier	哈丹吉尔杰库伦冰川	【挪威】
149	Haskell Glacier	哈斯克尔冰川	【南极洲】
150	Haut Glacier	浩特冰川	【瑞士】
151	Hayes Glacier	海斯冰川	【格陵兰】
152	Helheim Glacier	海尔海姆冰川	【格陵兰】
153	Hellstugubreen Glacier	海尔斯图古布林冰川	【挪威】
154	Helm Glacier	赫尔姆冰川	【加拿大】
155	Hercules Dome	赫尔克里士冰穹	【南极洲】
156	Hilda Glacier	希尔达冰川	【加拿大】
157	Hintereisferner Glacier	辛特艾斯费纳冰川	【奥地利】

续表

序号	英文名	中文名	所在地
158	Hispar Glacier	希斯帕尔冰川	【巴基斯坦】
159	Hlubeck Glacier	赫鲁贝克冰川	【南极洲】
160	Hodges Glacier	霍奇斯冰川	【格鲁吉亚】
161	Hofsjökull Glacier	霍夫斯加库尔冰川	【冰岛】
162	Honeycomb Glacier	蜂窝冰川	【美国】
163	Hooker Glacier	胡克冰川	【新西兰】
164	Hubbard Glacier	哈伯德冰川	【美国】
165	Humboldt Glacier	洪堡冰川	【格陵兰】
166	Hurd Glacier	赫德冰川	【南极洲】
167	Ice Worm Glacier	冰虫冰川	【美国】
168	Igdlugdlip Glacier	伊戈德鲁格德利普冰川	【格陵兰】
169	Ikertivaq Glacier	伊克提瓦格冰川	【格陵兰】
170	Inilchek Glacier	伊力切冰川	【吉尔吉斯斯坦】
171	Insukati Glacier	音苏盖提冰川	【中国】
172	Irenebreen Glacier	艾琳布林冰川	【挪威】
173	Irian Jaya Glacier	伊里安查亚冰川	【印度尼西亚】
174	Isbrecht Glacier	易斯布拉西特冰川	【南极洲】
175	Isfallsglaciären Glacier	易斯佛尔斯格拉西亚然冰川	【瑞典】
176	Ivory Glacier	艾沃瑞冰川	【新西兰】
177	Jakobshavn Glacier	雅各布冰川	【格陵兰】
178	Jamapa Glacier	哈马帕冰川	【墨西哥】
179	Jelbart Ice Shelf	杰尔巴特冰架	【南极洲】
180	John Evans Glacier	约翰埃文斯冰川	【加拿大】
181	Johnsons Glacier	约翰逊冰川	【南极洲】
182	Jones Ice Shelf	琼斯冰架	【南极洲】
183	Jorge Montt Glacier	豪尔赫蒙特冰川	【智利】
184	Juncal Norte Glacier	北洪卡尔峰冰川	【智利】

续表

序号	英文名	中文名	所在地
185	Kaffiφyra Glacier	卡非亚拉冰川	【挪威】
186	Kafni Glacier	卡夫尼冰川	【印度】
187	Kalabaland Glacier	卡拉巴朗德冰川	【印度】
188	Kaltamak Glacier	卡尔塔马克冰川	【中国】
189	Kanas Glacier	喀纳斯冰川	【中国】
190	Kangerdlugssuaq Glacier	康格尔得拉格斯瓦卡冰川	【格陵兰】
191	Kangwure Glacier	抗物热冰川	【中国】
192	Kannheiser Glacier	康海萨冰川	【南极洲】
193	Kaqin Glacier	恰青冰川	【中国】
194	Karayayilake Glacier	卡拉亚依拉克冰川	【中国】
195	Karola Glacier	卡若拉冰川	【中国】
196	Kennicott Glacier	肯尼科特冰川	【美国】
197	Kersten Glacier	克斯滕冰川	【坦桑尼亚】
198	Kexicar Glacier	科契喀尔冰川	【中国】
199	Khumbu Glacier	孔布冰川	【尼泊尔】
200	Kilimanjaro Glaciers	乞力马扎罗山冰川群	【坦桑尼亚】
201	Koeldukvislarjökull Glacier	柯尔杜克维斯拉加库尔冰川	【冰岛】
202	Koettlitz Glacier	柯尔特里茨冰川	【南极洲】
203	Kolka Glacier	科尔卡冰川	【俄罗斯】
204	Kong Oscar Glacier	港奥斯卡冰川	【格陵兰】
205	Kongsvegen Glacier	康斯韦根冰川	【挪威】
206	Koryto Glacie	科尔伊特冰川	【俄罗斯】
207	Krapf Glacier	克拉普夫冰川	【肯尼亚】
208	Kyagar Glacier	克亚吉尔冰川	【中国】
209	La Conejera (Sta Isabel) Glacier	拉康内吉拉(斯塔伊斯贝尔)冰川	【哥伦比亚】
210	La Paloma Glacier	拉帕洛马冰川	【智利】

续表

序号	英文名	中文名	所在地
211	Laigu Glacier	来古冰川	【中国】
212	Lambert Glacier	兰伯特冰川	【南极洲】
213	Langfjordjoekul Glacier	朗福吉德加库尔冰川	【挪威】
214	Langjökull Southern Dome	朗加库尔南冰穹	【冰岛】
215	Langtang Glacier	朗塘冰川	【尼泊尔】
216	Lanong Glacier	拉弄冰川	【中国】
217	Larsen Ice Shelf	拉森冰架	【南极洲】
218	Lazarev Ice Shelf	拉扎列夫冰架	【南极洲】
219	LeConte Glacier	勒孔特冰川	【美国】
220	Lemon Creek Glacier	莱蒙河冰川	【美国】
221	Leones Glacier	利昂冰川	【智利】
222	Leverett Glacier	莱弗里特冰川	【格陵兰】
223	Leviy Aktru Glacier	列维阿卡特鲁冰川	【俄罗斯】
224	Levko Glacier	列夫卡冰川	【南极洲】
225	Lewis Glacier	刘易斯冰川	【肯尼亚】
226	Lirung Glacier	利龙冰川	【尼泊尔】
227	Little Penck Glacier	小彭克冰川	【坦桑尼亚】
228	Litz Glacier	利茨冰川	【南极洲】
229	Lobbia Glacier	罗比亚冰川	【意大利】
230	Long Glacier	朗冰川	【南极洲】
231	Lower Curtis Glacier	下柯蒂斯冰川	【美国】
232	Lower Grindelwald Glacier	下格林德瓦冰川	【瑞士】
233	Lucchitta Glacier	鲁奇塔冰川	【南极洲】
234	Lunga (Vedretta) Glacier	朗格(维德莱特)冰川	【意大利】
235	Lyman Glacier	莱曼冰川	【美国】
236	Lynch Glacier	林奇冰川	【美国】
237	Lys Glacier	里斯冰川	【意大利】

续表

序号	英文名	中文名	所在地
238	Madaccio Glacier	马达西奥冰川	【意大利】
239	Mahaffey Glacier	马哈菲冰川	【南极洲】
240	Maladeta Glacier	马拉得塔冰川	【西班牙】
241	Malan Ice Cap	马兰冰帽	【中国】
242	Malaspina Glacier	马拉斯皮纳冰川	【美国】
243	Malavalle Glacier	马拉维勒冰川	【意大利】
244	Maliy Aktru Glacier	玛丽依阿卡特鲁冰川	【俄罗斯】
245	Mandrone Glacier	曼德伦冰川	【意大利】
246	Marck Glacier	马克冰川	【南极洲】
247	Marinelli Glacier	马里奈丽冰川	【智利】
248	Marmaglaciaeren Glacier	马玛格拉西艾伦冰川	【瑞典】
249	Martial Este Glacier	马歇尔艾斯特冰川	【阿根廷】
250	Matanuska Glacier	马塔努斯卡冰川	【美国】
251	Matusevich Glacier	马图谢维奇冰川	【南极洲】
252	McCall Glacier	麦考尔冰川	【美国】
253	McCarty Glacier	麦卡蒂冰川	【南极洲】
254	Medvezhiy Glacier	梅德韦日冰川	【塔吉克斯坦】
255	Meighen Ice Cap	米恩冰帽	【加拿大】
256	Melang Glacier	明永冰川	【中国】
257	Mendenhall Glacier	门登霍尔冰川	【美国】
258	Meola Glacier	米奥拉冰川	【印度】
259	Mer de Glacier	梅冰川	【法国】
260	Mertz Glacier	默茨冰川	【南极洲】
261	Miage Glacier	米亚吉冰川	【意大利】
262	Miar Glacier	米亚尔冰川	【巴基斯坦】
263	Midtre Lovénbreen Glacier	米德特拉文伯林冰川	【挪威】
264	Midui Glacier	米堆冰川	【中国】

续表

序号	英文名	中文名	所在地
265	Milam Glacier	米拉姆冰川	【印度】
266	Mincer Glacier	明瑟冰川	【南极洲】
267	Mittivakkat Glacier	米的瓦卡特冰川	【格陵兰】
268	Montasio Glacier	蒙大西奥冰川	【意大利】
269	Monte Perdido Glacier	蒙特珀迪多冰川	【西班牙】
270	Monteratsch Glacier	莫特拉奇冰川	【瑞士】
271	Moore Glacier	穆尔冰川	【乌干达】
272	Morelli Glacier	莫雷利冰川	【南极洲】
273	Moreno Glacier	莫雷诺冰川	【阿根廷】
274	Mueller Glacier	米勒冰川	【新西兰】
275	Muir Glacier	缪尔冰川	【美国】
276	Mukkila Glacier	木基拉冰川	【印度】
277	Muldrow Glacier	马尔德罗冰川	【美国】
278	Müller Ice Shelf	穆勒冰架	【南极洲】
279	Murchison Glacier	默奇森冰川	【新西兰】
280	Muztag Glacier	木孜塔格冰川	【中国】
281	Muztagh Ata Glacier	慕士塔格冰川	【中国】
282	Myers Glacier	迈尔斯冰川	【南极洲】
283	Naimonányi Glacier	纳木那尼冰川	【中国】
284	Nef Glacier	内夫冰川	【智利】
285	Nellie Juan Glacier	内利胡安冰川	【美国】
286	Nelson Ice Cap	纳尔逊冰帽	【南极洲】
287	Ngozumpa Glacier	恩格宗帕冰川	【尼泊尔】
288	Nigardsbreen Glacier	尼加尔德斯伯林冰川	【挪威】
289	Ninnis Glacier	宁尼斯冰川	【南极】
290	Nioghalvfjerdsbrae Glacier	尼奥加弗杰德斯布拉冰川	【格陵兰】
291	Nisqually Glacier	尼斯阔利冰川	【美国】

续表

序号	英文名	中文名	所在地
292	Noijinkangsang Glacier	宁金岗桑冰川	【中国】
293	Noisy Creek Glacier	喧港冰川	【美国】
294	Nonsurveyed Glacier	南苏尔维耶得冰川	【格陵兰】
295	Nordenskiold Glacier	诺登斯基尔德冰川	【格陵兰】
296	North Klawatti Glacier	北卡拉瓦迪冰川	【美国】
297	Northey Glacier	诺西冰川	【肯尼亚】
298	Nunatakavsaup Glacier	奴那达卡乌素普冰川	【格陵兰】
299	Obruchev Glacier	欧布鲁杰夫冰川	【俄罗斯】
300	O'Higgins Glacier	欧希金斯冰川	【新西兰】
301	Onelli Glacier	万力冰川	【阿根廷】
302	Ossoue Glacier	欧苏维冰川	【法国】
303	Ostenfeld Glacier	欧斯坦菲尔德冰川	【格陵兰】
304	Panmah Glacier	潘玛冰川	【巴基斯坦】
305	Paron Glacier	帕朗冰川	【秘鲁】
306	Passu Glacier	帕苏冰川	【巴基斯坦】
307	Pasterze Glacier	巴斯特泽冰川	【奥地利】
308	Payne Glacier	佩恩冰川	【南极洲】
309	Pelter Glacier	佩尔特冰川	【南极洲】
310	Penck Glacier	彭克冰川	【坦桑尼亚】
311	Pendente Glacier	彭丹特冰川	【意大利】
312	Perad Glacier	皮拉德冰川	【印度】
313	Petermann Glacier	彼得曼冰川	【格陵兰】
314	Petrova Glacier	彼得罗瓦冰川	【吉尔吉斯斯坦】
315	Peyto Glacier	佩特冰川	【加拿大】
316	Pindari Glacier	品达理冰川	【印度】
317	Pine Island Glacier	派因艾兰冰川	【南极洲】
318	Pizol Glacier	皮错尔冰川	【瑞士】

续表

序号	英文名	中文名	所在地
319	Place Glacier	普莱斯冰川	【加拿大】
320	Potanin Glacier	波塔宁冰川	【蒙古】
321	Puruogangri Ice Field	普若岗日冰原	【中国】
322	Qiangyong Glacier	枪勇冰川	【中国】
323	Qirbulak Glacier	切尔布拉克冰川	【中国】
324	Qiyi Glacier	"七一"冰川	【中国】
325	Qori Kalis Glacier	居里卡利斯冰川	【秘鲁】
326	Quar Ice Shelf	奎尔冰架	【南极洲】
327	Quelccaya Ice Cap	奎卡亚冰帽	【秘鲁】
328	Rabots Glacier	拉伯茨冰川	【瑞典】
329	Rainbow Glacier	彩虹冰川	【美国】
330	Ratzel Glacier	拉杰尔冰川	【坦桑尼亚】
331	Rebmann Glacier	瑞布曼冰川	【坦桑尼亚】
332	Rexford Glacier	雷克斯福德冰川	【南极洲】
333	Rhone Glacier	龙冰川	【瑞士】
334	Rignot Glacier	里格诺特冰川	【南极洲】
335	Riiser-Larsen Ice Shelf	里塞—拉森冰架	【南极洲】
336	Rinks Glacier	林克斯冰川	【格陵兰】
337	Río Blanco Glacier	里约布兰科冰川	【阿根廷】
338	Ritacuba Negro (La Cocuy) Glacier	里特古巴内格罗（拉科甲）冰川	【哥伦比亚】
339	Riukojietna Glacier	瑞克杰塔纳冰川	【瑞典】
340	Robbins Glacier	罗宾斯冰川	【南极洲】
341	Rochray Glacier	洛基来冰川	【南极洲】
342	Rongbuk Glacier	绒布冰川	【中国】
343	Ronne Ice Shelf	罗纳冰架	【南极洲】
344	Rosanova Glacier	罗萨拿法冰川	【南极洲】

续表

序号	英文名	中文名	所在地
345	Rundvassbreen Glacier	朗德瓦斯伯林冰川	【挪威】
346	Russell Glacier	罗塞尔冰川	【格陵兰】
347	Rutor Glacier	鲁特冰川	【瑞士】
348	Ryder Glacier	赖德冰川	【格陵兰】
349	Saint Sorlin Glacier	圣索林冰川	【法国】
350	Sandalee Glacier	桑达里冰川	【美国】
351	Sarennes Glacier	萨仁内斯冰川	【法国】
352	Sarpo Laggo Glacier	萨帕拉格冰川	【巴基斯坦】
353	Saskatchewan Glacier	萨斯喀彻温冰川	【加拿大】
354	Savage Glacier	萨维奇冰川	【南极洲】
355	Savoia Glacier	萨瓦冰川	【乌干达】
356	Scandinavian Ice Sheet	斯堪的纳维亚古冰盖	【北欧】
357	Sentinel Glacier	森蒂纳尔冰川	【加拿大】
358	Sermersuaq (Humboldt) Glacier	瑟莫斯瓦克(洪堡)冰川	【格陵兰】
359	Sessums Glacier	塞萨姆斯冰川	【南极洲】
360	Shackleton Ice Shelf	沙克尔顿冰架	【南极洲】
361	Sholes Glacier	肖尔斯冰川	【美国】
362	Shumsky Glacier	舒姆斯基冰川	【哈萨克斯坦】
363	Shunkalpa (Ralam) Glacier	顺卡尔帕(拉兰姆)冰川	【印度】
364	Siachen Glacier(Rose Glacier)	锡亚琴冰川(玫瑰冰川)	【克什米尔】
365	Sikorski Glacier	西科尔斯基冰川	【南极洲】
366	Silver Glacier	银冰川	【美国】
367	Silvretta Glacier	希尔瓦雷塔冰川	【瑞士】
368	Siple Dome	赛普尔冰穹	【南极洲】
369	Slava Ice Shelf	斯拉法冰架	【南极洲】
370	Small River Glacier	小河冰川	【加拿大】

续表

序号	英文名	中文名	所在地
371	Sofiyskiy Glacier	索菲斯基冰川	【俄罗斯】
372	Soler Glacier	索莱尔冰川	【智利】
373	Sonapani Glacier	所纳帕尼冰川	【印度】
374	Sonnblick Glacier	松布利克冰川	【奥地利】
375	Sorbreen Glacier	索尔布林冰川	【挪威】
376	South Cascade Glacier	南喀斯卡特冰川	【美国】
377	Southeast Glacier	东南冰川	【格陵兰】
378	Speel Glacier	斯皮尔冰川	【美国】
379	Speke Glacier	斯皮克冰川	【乌干达】
380	Stapleton Glacier	斯特普尔顿冰川	【南极洲】
381	Steenstrup Glacier	斯廷斯特鲁普冰川	【格陵兰】
382	Steffen Glacier	斯蒂芬冰川	【智利】
383	Storbreen Glacier	斯托伯林冰川	【挪威】
384	Storglaciaeren Glacier	斯托格雷希尔仑冰川	【瑞典】
385	Storglombreen Glacier	斯托格拉姆伯林冰川	【挪威】
386	Storstrømmen Glacier	斯托斯特蒙冰川	【格陵兰】
387	Susitna Glacier	苏西特纳冰川	【美国】
388	Svelgjabreen Glacier	斯瓦尔格加伯林冰川	【挪威】
389	Swinburne Ice Shelf	斯威伯恩冰架	【南极洲】
390	Tailan Glacier	台兰冰川	【中国】
391	Taku Glacier	塔库冰川	【美国】
392	Talos Dome	塔罗斯冰穹	【南极洲】
393	Tarfalaglaciaeren Glacier	特拉法拉格拉希尔仑冰川	【瑞典】
394	Tasman Glacier	塔斯曼冰川	【新西兰】
395	Taylor Dome	泰勒冰穹	【南极洲】
396	Telamukli Glacier	特拉木坎力冰川	【中国】
397	Thomson Glacier	汤姆森冰川	【南极洲】

续表

序号	英文名	中文名	所在地
398	Tikke Glacier	提克冰川	【加拿大】
399	Tipra Bank Glacier	迪普拉邦克冰川	【印度】
400	Trango Glacier	特朗格冰川	【巴基斯坦】
401	Triolet Glacier	特里奥莱冰川	【意大利】
402	Tungnaarjökull Glacier	唐纳加库尔冰川	【冰岛】
403	Tuyuksu Glacier	图尤克苏冰川	【哈萨克斯坦】
404	Uhlig Glacier	乌利希冰川	【坦桑尼亚】
405	Upernavik Glacier	乌佩纳维克冰川	【格陵兰】
406	Upper Seward Glacier	上西沃德冰川	【加拿大】
407	Upsala Glacier	乌普萨拉冰川	【阿根廷】
408	Van de Water Glacier	范德沃特冰川	【印度尼西亚】
409	Variegated Glacier	花冰川	【美国】
410	Vatnajökull Glacier	瓦特那加库尔冰川	【冰岛】
411	Vedretta Piana Glacier	瓦德莱特皮亚娜冰川	【意大利】
412	Velasco Glacier	韦拉斯科冰川	【南极洲】
413	Vernagtferner Glacier	瓦纳克费纳冰川	【奥地利】
414	Viedma Glacier	别德马冰川	【阿根廷】
415	Vodopadniy Glacier	瓦德帕德尼冰川	【俄罗斯】
416	Volta Glacier	沃尔塔冰川	【新西兰】
417	Waldemar Glacier	瓦尔德马冰川	【挪威】
418	Waldemarbreen Glacier	瓦尔德马布林冰川	【挪威】
419	Walk Glacier	沃克冰川	【南极洲】
420	Warr Glacier	沃尔冰川	【南极洲】
421	Wedgemount Glacier	楔子山冰川	【加拿大】
422	West Fork Glacier	西福克冰川	【美国】
423	White Glacier	白冰川	【加拿大】
424	Wollaston Glacier	沃拉斯顿冰川	【印度尼西亚】

续表

序号	英文名	中文名	所在地
425	Wolverine Glacier	沃尔弗林冰川	【美国】
426	Worthington Glacier	沃辛顿冰川	【美国】
427	Xinqingfeng Glacier	新青峰冰川	【中国】
428	Yanamarey Glacier	亚娜马雷冰川	【秘鲁】
429	Yanert Glacier	亚讷特冰川	【美国】
430	Yangbulak Glacier	洋布拉克冰川	【中国】
431	Yawning Glacier	亚宁冰川	【美国】
432	Yehelong Glacier	耶和龙冰川	【中国】
433	Zemu Glacier	则木冰川	【印度】
434	Zepu Glacier	则普冰川	【中国】
435	Zinberg Glacier	津伯格冰川	【南极洲】
436	Zongo Glacier	宗戈冰川	【玻利维亚】
437	Zuoqiupu Glacier	作求普冰川	【中国】

附录3　　　　国内外主要期刊

(1) 英文期刊

序号	英文名称	中文名称
1	Acta Geologica Polonica	波兰地质学报【波兰】
2	Acta Meteorologica Sinica	气象学报【中国】
3	Advances in Atmospheric Sciences	大气科学进展【中国】
4	Advances in Climate Change Research	气候变化研究进展【中国】
5	Advances in Space Research	空间研究进展【法国】
6	American Scientist	美国科学家【美国】
7	Annales Geophysicae	地球物理学集刊【德国】
8	Annals of Glaciology	冰川学集刊【英国】
9	Annals of the Association of American Geographers	美国地理学家协会年鉴【美国】
10	Annual Review of Earth and Planetary Sciences	地球和行星科学年度回顾【美国】
11	Antarctic Science	南极科学【英国】
12	Arctic	北极【加拿大】
13	Arctic, Antarctic, and Alpine Research	北极、南极和高山研究【美国】
14	Atmosphere-Ocean	大气－海洋【加拿大】
15	Atmospheric and Oceanic Science Letters	大气海洋科学通讯【中国】
16	Atmospheric Chemistry and Physics	大气化学与物理【德国/EGU】
17	Atmospheric Environment	大气环境【英国】
18	Atmospheric Research	大气研究【荷兰】
19	Boreas	北风【挪威】
20	Bulletin of American Meteorological Society	美国气象学会通报【美国】
21	Bulletin of Australian Meteorological and Oceanographic Society	澳大利亚气象与海洋学会通报【澳大利亚】
22	Bulletin of Glacier Research	冰川研究通报【日本】

续表

序号	英文名称	中文名称
23	Bulletin of the Seismological Society of America	美国地震学会通报【美国】
24	Canadian Geotechnical Journal	加拿大岩土工程学报【加拿大】
25	Canadian Journal of Forest Research	加拿大森林研究杂志【加拿大】
26	Chemical Geology	化学地质【荷兰】
27	Chinese Journal of Geophysics	地球物理学报【中国】
28	Chinese Journal of Polar Science	极地科学学报【中国】
29	Chinese Science Bulletin	科学通报【中国】
30	Climate Dynamics	气候动力学【美国】
31	Climate Research	气候研究【德国】
32	Climatic Change	气候变化【荷兰】
33	Cold Region Science and Technology	寒区科学与技术【荷兰】
34	Earth and Planetary Science Letters	地球和行星科学通讯【荷兰】
35	Earth Interactions	地球表面相互作用杂志【美国】
36	Earth Surface Processes and Landforms	地表过程与地貌【英国】
37	Earth, planets, and space	地球、行星与太空【日本】
38	Earth-Science Reviews	地球科学评论【荷兰】
39	Ecological Economics	生态经济学【美国】
40	Ecological Indicators	生态指标【美国】
41	Ecological Modeling	生态模拟【丹麦】
42	Ecological Monographs	生态学集刊【美国】
43	Environmental Research Letters	环境研究通讯【英国】
44	EOS, Transactions American Geophysical Union	美国地球物理学联合会会刊【美国】
45	Geochemistry, Geophysics, Geosystems	地球化学、地球物理和地球系统【美国】
46	Geological Society of America Bulletin	美国地质学会通报【美国】
47	Geology	地质学【美国】

续表

序号	英文名称	中文名称
48	Geomorphology	地貌学【荷兰】
49	Geophysical and Astrophysical Fluid Dynamics	地球物理与天体物理流体动力学【英国】
50	Geophysical Journal International	国际地球物理杂志【英国】
51	Geophysical Research Letters	地球物理学研究通讯【美国】
52	Global and Planetary Change	地球与行星变化【荷兰】
53	Global Biogeochemical Cycles	全球生物地球化学循环【美国】
54	Global Change Biology	全球变化生物学【英国】
55	Himalayan Journal of Sciences	喜马拉雅科学杂志【尼泊尔】
56	Hydrogeology Journal	水文地质学杂志【德国】
57	Hydrological Processes	水文过程【英国】
58	Hydrological Sciences Journal	水文科学杂志【英国】
59	Hydrology and Earth System Sciences	水文学与地球系统科学【德国】
60	Ice	冰杂志【英国】
61	IEEE Transaction on Geoscience and Remote Sensing	电气电子工程师协会会刊—地球科学与遥感【美国】
62	International Journal of Climatology	国际气候学杂志【英国】
63	International Journal of Remote Sensing	国际遥感学杂志【英国】
64	Journal of Applied Geophysics	应用地球物理学杂志【荷兰】
65	Journal of Applied Meteorology and Climatology	应用气象气候杂志【美国】
66	Journal of Arid Land Science	旱区科学研究杂志【中国】
67	Journal of Asian Earth Sciences	亚洲地球科学【英国】
68	Journal of Atmospheric and Oceanic Technology	大气与海洋技术杂志【美国】
69	Journal of Atmospheric and Solar-Terrestrial Physics	大气物理学与日地物理学杂志【美国】
70	Journal of Climate	气候杂志【美国】

续表

序号	英文名称	中文名称
71	Journal of Cold Regions Engineering	寒区工程杂志【美国】
72	Journal of Geodynamics	地球动力学杂志【荷兰】
73	Journal of Geographic Science	地理科学杂志【中国】
74	Journal of Geophysical Research	地球物理研究【美国】
75	Journal of Geophysical Research-Atmospheres	地球物理研究杂志－大气【美国】
76	Journal of Geophysical Research-Biogeosciences	地球物理研究杂志－生物地球科学【美国】
77	Journal of Geophysical Research-Oceans	地球物理研究杂志－海洋【美国】
78	Journal of Geophysical Research-Earth Surface	地球物理研究杂志－地球表面【美国】
79	Journal of Geophysical Research-Solid Earth	地球物理研究杂志－固体地球【美国】
80	Journal of Geophysical Research-Space Physics	地球物理研究－空间物理【美国】
81	Journal of Glaciology	冰川学杂志【英国】
82	Journal of Hydrology	水文学杂志【荷兰】
83	Journal of Hydrometeorology	水文气象学杂志【美国】
84	Journal of Meteorological Society of Japan	日本气象学会杂志【日本】
85	Journal of Mountain Science	山地科学杂志【中国】
86	Journal of Physical Oceanography	物理海洋学杂志【美国】
87	Journal of Quaternary Science	第四纪科学【英国】
88	Journal of the Atmospheric Sciences	大气科学杂志【美国】
89	Marine Geology	海洋地质学【荷兰】
90	Meteorological Monographs	气象学集刊【美国】
91	Meteorology and Atmospheric Physics	气象学与大气物理学【德国】
92	Monthly Weather Review	每月天气评论【美国】

续表

序号	英文名称	中文名称
93	Mountain Research and Development	山地研究与发展【美国】
94	Nature	自然【英国】
95	Nature Climate Change	自然气候变化【美国】
96	Nonlinear Processes in Geophysics	地球物理的非线性过程【德国】
97	Norwegian Journal of Geography	挪威地理杂志【挪威】
98	Palaeogeography, Palaeoclimatology, Palaeoecology (PPP)	古地理、古气候和古生态学(三古杂志)【荷兰】
99	Paleoceanography	古海洋学【美国】
100	Permafrost and Periglacial Processes	多年冻土与冰缘过程【美国】
101	Physical Geography	自然地理【美国】
102	Physics of the Earth and Planetary Interiors	地球与行星内部物理学【荷兰】
103	Planetary and Space Science	行星与空间科学【美国】
104	Polar Biology	极地生物学【美国】
105	Polar Geography	极地地理学【英国】
106	Polar Record	极地记录【美国】
107	Polar Research	极地研究【挪威】
108	Progress in Natural Sciences	自然科学进展【中国】
109	Progress in Physical Geography	自然地理学进展【英国】
110	Pure and Applied Geophysics	理论与应用地球物理学【瑞士】
111	Quartenary International	国际第四纪杂志【英国】
112	Quarterly Journal of the Royal Meteorological Society	英国皇家气象协会季刊【英国】
113	Quaternary Geochronology	第四纪地球年代学【英国】
114	Quaternary Research	第四纪研究【美国】
115	Quaternary Science Reviews	第四纪科学评论【英国】
116	Radio Science	无线电科学【美国】
117	Remote Sensing of Environment	环境遥感【美国】

续表

序号	英文名称	中文名称
118	Reviews of Geophysics	地球物理学评论【美国】
119	Science	科学【美国】
120	Sciences in China Series D—Earth Sciences	中国科学 D 辑—地球科学【中国】
121	Sciences in Cold and Arid Regions	寒旱区科学【中国】
122	Scientific American	科学美国人【美国】
123	Space Science Reviews	空间科学评论【荷兰】
124	Tellus Series A—Dynamic Meteorology and Oceanography	大地系列 A 辑—动力气象和海洋【瑞典】
125	Tellus Series B—Chemical and Physical Meteorology	大地系列 B 辑—化学和物理气象学【瑞典】
126	The Cryosphere	冰冻圈【德国】
127	The Holocene	全新世【加拿大】
128	Theoretical and Applied Climatology	理论和应用气候学【美国】
129	Water Resource Research	水资源研究【美国】
130	Weather	天气【美国】
131	Weather and Forecasting	天气预报【美国】
132	Weather, Climate, and Society	气象、气候与社会【美国】

(2)中文期刊

序号	中文名称	英文名称
1	北方交通	Northern Communications
2	北京大学学报(自然科学版)	Acta Scientiarum Naturalium Universitatis Pekinensis (ASNUP)
3	北京师范大学学报(自然科学版)	Journal of Beijing Normal University (Natural Science)
4	冰川冻土	Journal of Glaciology and Geocryology
5	渤海大学学报(自然科学版)	Journal of Bohai University (Natural Science Edition)
6	草业科学	Acta Prataculturae Sinica
7	测绘科技情报	Science and Technology Information of Surveying and Mapping
8	测绘科学	Science of Surveying and Mapping
9	长春大学学报	Journal of Changchun University
10	长江流域资源与环境	Resources and Environment in the Yangtze Basin
11	大连海事大学学报	Journal of Dalian Maritime University
12	大连理工大学学报	Journal of Dalian University of Technology
13	大气科学	Chinese Journal of Atmospheric Sciences
14	大气科学学报(原南京气象学院学报)	Transaction of Atmospheric Sciences
15	低温建筑技术	Low Temperature Architecture Technology
16	低温与超导	Cryogenics and Superconductivity
17	地理科学	Scientia Geographica Sinica
18	地理空间信息	Geospatial Information
19	地理学报	Acta Geographica Sinica
20	地理研究	Geographical Research
21	地理与地理信息科学	Geography and Geo-Information Science
22	地球科学	Earth Science
23	地球科学进展	Advances in Earth Science
24	地球科学与环境学报	Journal of Earth Sciences and Environment

续表

序号	中文名称	英文名称
25	地球物理学报	Chinese Journal of Geophysics
26	地球信息科学学报	Geo-Information Science
27	地球学报	Acta Geoscientia Sinica
28	地下工程与隧道	Underground Engineering and Tunnels
29	地下空间与工程学报	Chinese Journal of Underground Space and Engineering
30	地学前缘	Earth Science Frontiers
31	地域研究与开发	Areal Research and Development
32	地质科学	Chinese Journal of Geology
33	地质评论	Geological Review
34	地质通报	Geological Bulletin of China
35	地质学报	Acta Geologica Sinica
36	第四纪研究	Quaternary Sciences
37	干旱气象	Journal of Arid Meteorology
38	干旱区地理	Arid Land Geography
39	干旱区研究	Arid Zone Research
40	干旱区资源与环境	Journal of Arid Land Resources and Environment
41	高原气象	Plateau Meteorology
42	工程地质学报	Journal of Engineering Geology
43	工程勘察	Geotechnical Investigation & Surveying
44	工程力学	Engineering Mechanics
45	工程与试验	Engineering and Test
46	公路	Highway
47	公路交通科技	Journal of Highway and Transportation Research and Development
48	管道技术与设备	Pipeline Technique and Equipment
49	光谱学与光谱分析	Spectroscopy and Spectral Analysis
50	国土与自然资源研究	Territory and Natural Resources Study

续表

序号	中文名称	英文名称
51	国土资源遥感	Remote Sensing for Land and Resources
52	国外油田工程	Foreign Oilfield Engineering
53	海洋科学	Marine Sciences
54	海洋世界	Ocean World
55	黑龙江大学自然科学学报	Journal of Natural Science of Heilongjiang University
56	湖泊科学	Journal of Lake Sciences
57	环境保护	Environmental Protection
58	环境化学	Environmental Chemistry
59	环境科学	Environmental Science
60	环境科学学报	Acta Scientiae Circumstantiae
61	环境科学研究	Research of Environmental Sciences
62	环境科学与管理	Environmental Science and Management
63	环境科学与技术	Environmental Science and Technology
64	吉林大学学报（地球科学版）	Journal of Jilin University (Earth Science Edition)
65	极地研究	Chinese Journal of Polar Research
66	科技创新导报	Science and Technology Innovation Herald
67	科技信息	Science and Technology Information
68	科学通报	Chinese Science Bulletin
69	兰州大学学报（自然科学版）	Journal of Lanzhou University (Natural Sciences)
70	冷藏技术	Cold Storage Technology
71	路基工程	Subgrade Engineering
72	南京大学学报（自然科学版）	Journal of Nanjing University (Natural Sciences)
73	南京师大学报（自然科学版）	Journal of Nanjing Normal University (Natural Science Edition)
74	气候变化研究进展	Advances in Climate Change Research

续表

序号	中文名称	英文名称
75	气候与环境研究	Climatic and Environmental Research
76	气象	Meteorological Monthly
77	气象科技	Meteorological Science and Technology
78	气象科学	Scientia Meteorologica Sinica
79	气象学报	Acta Meteorologica Sinica
80	气象与环境学报	Journal of Meteorology and Environment
81	沙漠与绿洲气象	Desert and Oasis Meteorology
82	山地学报	Journal of Mountain Science
83	生态经济	Ecological Economy
84	生态学报	Acta Ecologica Sinica
85	施工技术	Construction Technology
86	水科学进展	Advances in Water Science
87	水科学与工程技术	Water Sciences and Engineering Technology
88	水利与建筑工程学报	Journal of Water Resources and Architectural Engineering
89	水文	Journal of China Hydrology
90	水文地质工程地质	Hydrogeology and Engineering Geology
91	水资源与水工程学报	Journal of Water Resources and Water Engineering
92	隧道建设	Tunnel Construction
93	探矿工程	Exploration Engineering
94	铁道工程学报	Journal of Railway Engineering Society
95	铁道建筑技术	Railway Construction Technology
96	铁道勘察	Railway Investigation and Surveying
97	铁道科学与工程学报	Journal of Railway Science and Engineering
98	铁道学报	Journal of the China Railway Society
99	土木工程学报	China Civil Engineering Journal
100	土壤	Soils
101	土壤通报	Chinese Journal of Soil Science

续表

序号	中文名称	英文名称
102	武汉大学学报(信息科学版)	Geomatics and Information Science of Wuhan Uiversity
103	物理学报	Acta Physica Sinica
104	西部资源	Western Resources
105	现代地质	Geoscience
106	岩石力学与工程学报	Chinese Journal of Rock Mechanics and Engineering
107	岩土工程界	Geotechnical Engineering World
108	岩土工程学报	Chinese Journal of Geotechnical Engineering
109	岩土力学	Rock and Soil Mechanics
110	遥感技术与应用	Remote Sensing Technology and Applicaion
111	遥感信息	Remote Sensing Information
112	遥感学报	Journal of Remote Sensing
113	应用气象学报	Journal of Applied Meteorological Science
114	应用生态学报	Chinese Journal of Applied Ecology
115	油气田地面工程	Oil-Gasfield Surface Engineering
116	灾害学	Journal of Catastrophology
117	植物生态学报	Acta Phytoecologica Sinica
118	制冷与空调	Refrigeration and Air-Conditioning
119	中国地理与资源文摘	Chinese Geographical and Resources Abstracts
120	中国公路学报	China Journal of Highway and Transport
121	中国海洋大学学报(自然科学版)	Periodical of Ocean University of China (Natural Sciences)
122	中国环境科学	China Environmental Science
123	中国科技期刊研究	Chinese Journal of Scientific and Technical Periodicals
124	中国科学D辑:地球科学	Science in China Series D: Earth Sciences
125	中国农业科学	Scientia Agricultura Sinica
126	中国沙漠	Journal of Desert Research
127	中国铁道科学	China Railway Science

续表

序号	中文名称	英文名称
128	资源环境与工程	Resources Environment and Engineering
129	资源科学	Resources Science
130	自然科学进展	Progress in Natural Science
131	自然杂志	Chinese Journal of Nature
132	自然灾害学报	Journal of Natural Disasters
133	自然资源学报	Journal of Natural Resources

附录 4　　　国内外主要研究机构

(1) 国际冰冻圈研究机构

序号	英文名	中文名	所属国家
1	Ahmed Yasawi International Kazakh-Turkish University, Turkestan, Kazakhstan	国际哈萨克—土耳其大学	【哈萨克斯坦】
2	Alaska Climate Research Center	阿拉斯加气候研究中心	【美国】
3	Alfred Wegener Institute for Polar and Marine Research	阿尔弗雷德极地和海洋研究所	【德国】
4	Almaty Institute of Power Engineering and Telecommunications	阿拉木图电力工程与电信研究所	【哈萨克斯坦】
5	Applied Physics Laboratory Polar Science Center, University of Washington Seattle	华盛顿大学应用物理实验室极地科学中心	【美国】
6	Arctic and Antarctic Research Institute	北极和南极研究所	【俄罗斯】
7	Arctic and Antarctic Research Institute, Russian Federation Sevice for Hydrometeorology and Environmental Monitoring	俄罗斯联邦水文气象和环境监测服务中心南北极研究所	【俄罗斯】
8	Arctic Center, University of Lapland	拉普兰大学北极中心	【芬兰】
9	Arctic Centre	北极中心	【芬兰】
10	Arctic Geology Department, the University Centre in Svalbard	斯瓦尔巴德群岛大学中心北极地质系	【挪威】
11	Arctic Institute of North America	北美北极研究所	【加拿大】
12	Arctic Institute of North America, University of Calgary	卡尔加里大学北美北极研究所	【加拿大】
13	Arctic Research Center of Finnish Meteorological Institute	芬兰气象研究所北极研究中心	【芬兰】
14	Argentine Antarctic Institute	阿根廷南极研究所	【阿根廷】
15	Australian Antarctic Division and the Antarctic Climate and Ecosystems CRC	澳大利亚南极局南极气候与生态系统合作研究中心	【澳大利亚】

续表

序号	英文名	中文名	所属国家
16	Australian Antarctic Divison	澳大利亚南极局	【奥大利亚】
17	Brazil's National Institute for Cryospheric Science	巴西冰冻圈科学国立研究所	【巴西】
18	Bristol Glaciology Centre, School of Geographical Sciences, University of Bristol	布里斯托尔大学地理科学学院布里斯托尔冰川中心	【英国】
19	British Antarctic Survey (BAS)	英国南极局	【英国】
20	Byrd Polar Research Center, Ohio State University	俄亥俄州立大学伯德极地研究中心	【美国】
21	Canadian Circum-Polar Institute	加拿大环北极研究所	【加拿大】
22	Canadian Ice Service	加拿大冰情服务处	【加拿大】
23	Center for Geophysical Studies of Ice and Climate, St. Olaf College	圣奥拉夫学院冰和气候物理研究中心	【美国】
24	Center for Remote Sensing of Ice Sheets (CReSIS)	冰盖遥感中心	【美国】
25	Center for Snow & Avalanche Studies	美国雪崩研究中心	【美国】
26	Central Asia Institute of Applied Geosciences	中亚应用地学研究所	【吉尔吉斯斯坦】
27	Centre for Climate and Cryosphere, University of Innsbruck	因斯布鲁克大学气候和冰冻圈中心	【奥地利】
28	Centre for Glaciology, University of Wales	威尔士大学冰川学中心	【英国】
29	Centre for Ice and Climate, University of Copenhagen	哥本哈根大学冰与气候中心	【丹麦】
30	Chile Laboratoria de Glaciologia, Universidad de Magallanes	麦哲伦大学冰川实验室	【智利】
31	Chilean Antarctic Institute	智利南极研究所	【智利】
32	Climate Change Coordination Center	气候变化协调中心	【哈萨克斯坦】
33	Climate Change Institute, University of Maine	缅因大学气候变化研究所	【美国】

续表

序号	英文名	中文名	所属国家
34	Climate Change Research Center, University of New Hampshire	新罕布什尔大学气候变化研究中心	【美国】
35	Climatic Research Unit, University of East Anglia	东英吉利大学气候研究中心	【英国】
36	Cold Regions Research Center, Wilfrid Laurier University	威尔弗里德·劳里埃大学寒区研究中心	【加拿大】
37	College of Military Engineering	军事工程学院	【印度】
38	Cooperative Institute for Arctic Research (CIFAR), University of Alaska	阿拉斯加大学北极研究合作学院	【美国】
39	Cooperative Institute for Research in Environmental Science, University of Colorado at Boulder	科罗拉多大学博尔德分校环境科学研究合作学院	【美国】
40	Cryospheric and Polar Processes Division, University of Colorado	科罗拉多大学冰冻圈和极地过程系	【美国】
41	Danish Meteorological Institute	丹麦气象研究所	【丹麦】
42	Danish National Space Center, Technical University of Denmark	丹麦技术大学丹麦国家航天中心	【丹麦】
43	Department of Arctic Geophysics, The University Centre in Svalbard (UNIS)	斯瓦尔巴特大学中心北极地球物理系	【挪威】
44	Department of Chemistry, University of Florence	佛罗伦萨大学化学系	【意大利】
45	Department of Earth and Ecosystem Sciences, Lünd University	隆德大学地球和生态系统科学系	【瑞典】
46	Department of Earth Sciences, Uppsala University	乌普萨拉大学地球科学系	【瑞典】
47	Department of Ecological Engineering, Kyrgyzstan-Türkey Manas University	玛纳斯大学生态工程系	【吉尔吉斯斯坦】
48	Department of Geography and Environmental management, University of Waterloo	滑铁卢大学地理与环境科学系	【加拿大】
49	Department of Geography, ETH	瑞士联邦理工学院地理系	【瑞士】

续表

序号	英文名	中文名	所属国家
50	Department of Geography, Pennsylvania State University	宾夕法尼亚州立大学地理系	【美国】
51	Department of Geography, The George Washington University	乔治华盛顿大学地理系	【美国】
52	Department of Geography, University of Calgary	卡尔加里大学地理系	【加拿大】
53	Department of Geography, University of Cambridge	剑桥大学地理系	【英国】
54	Department of Geography, University of Delaware	特拉华大学地理系	【美国】
55	Department of Geography, University of Manchester	曼彻斯特大学地理系	【英国】
56	Department of Geography, University of Sheffield	谢菲尔德大学地理系	【英国】
57	Department of Geography, University of Zurich	苏黎世大学地理系	【瑞士】
58	Department of Meteorology, Stockholm University	斯德哥尔摩大学气象科学系	【瑞典】
59	Department of Physical Geography and Quaternary Geology, Stockholm University	斯德哥尔摩大学自然地理和第四纪地质学系	【瑞典】
60	Desert Reserch Institute	沙漠研究所	【美国】
61	Ecosystem Center, Marine Biological Laboratory	海洋生物实验室生态系统中心	【美国】
62	Ecuador Antarctic Institute	厄瓜多尔南极研究所	【厄瓜多尔】
63	Geological and Planetary Sciences, California Institute of Technology	加州理工学院地质与行星科学系	【美国】
64	Geological Survey of Brazil	巴西地质调查局	【巴西】
65	Geological Survey of Denmark and Greenland (GEUS)	丹麦和格陵兰地质调查局	【丹麦】
66	Geological Survey of Peru	秘鲁地质调查局	【秘鲁】

续表

序号	英文名	中文名	所属国家
67	Geophysical Institute, University of Alaska Fairbanks	阿拉斯加大学费尔班克斯分校地球物理研究所	【美国】
68	Glaciology and Climate Change, Centro de Estudios Científicos (Center for Scientific Studies) (CECs)	智利科学研究中心冰川与气候研究所	【智利】
69	Glaciology Group, Department of Geophysics and Astronomy, University of British Columbia	大不列颠哥伦比亚大学地球物理学和天文学系冰川学组	【加拿大】
70	Global Hydrology and Climate Center, NASA	美国航空航天局全球水文和气候中心	【美国】
71	Indian Institute of Science	印度科学研究所	【印度】
72	Indian Institute of Technology	印度技术研究所	【印度】
73	Institute for Atmospheric and Climate Science, Swiss Federal Institute of Technology, ETH	瑞士联邦理工学院大气和气候学研究所	【瑞士】
74	Institute for Environmental Physics, University of Heidelberg	海德堡大学环境物理学研究所	【德国】
75	Institute for Geophysics, University of Texas at Austin	德克萨斯大学奥斯汀分校地球物理研究所	【美国】
76	Institute for Hydrogeology and Hydrophysics, Academy of Sciences of Kazakhstan	哈萨克斯坦科学院水文地质与水文物理研究所	【哈萨克斯坦】
77	Institute for Hydrospheric-Atmospheric Sciences, Nagoya University	名古屋大学水文气象研究所	【日本】
78	Institute for Meteorology and Climate Research, Forschungszentrum Karlsruhe, University of Karlsruhe	卡尔斯鲁厄大学气象和气候研究所	【德国】
79	Institute of Applied Physics, University of Bern	伯尔尼大学应用物理研究所	【瑞士】
80	Institute of Arctic and Alpine Research (INSTAAR), University of Colorado	科罗拉多大学北极与高山研究所	【美国】
81	Institute of Arctic and Antarctic Research, Russian Federal	俄罗斯联邦南北极研究所	【俄罗斯】

续表

序号	英文名	中文名	所属国家
82	Institute of Arctic Biology, University of Alaska Fairbanks	阿拉斯加大学费尔班克斯分校北极生物研究所	【美国】
83	Institute of Environmental Engineering, ETH	瑞士联邦理工学院环境工程研究所	【瑞士】
84	Institute of Geography and Geology, University of Copenhagen	哥本哈根大学地理与地质研究所	【丹麦】
85	Institute of Geography, Russian Academy of Sciences	俄罗斯科学院地理研究所	【俄罗斯】
86	Institute of Geology, University of Bonn	波恩大学地质研究所	【德国】
87	Institute of Geophysics, University of Oslo	奥斯陆大学地球物理研究所	【挪威】
88	Institute of Geosciences, University of Graz	格拉茨大学地球科学研究所	【奥地利】
89	Institute of Low Temperature Science, Hokkaido University	北海道大学低温科学研究所	【日本】
90	Institute of Space and Planetary Astrophysics (ISPA), University of Karachi	卡拉奇大学空间与行星天气物理学研究所	【巴基斯坦】
91	Institute of the Earth Cryosphere, Siberian Branch of Russian Academy of Sciences	俄罗斯科学院西伯利亚分院地球冰冻圈研究所	【俄罗斯】
92	Institute of the Observational Research for Global Change (IORGC)	日本全球变化观测研究所	【日本】
93	Institute of Water Problems and Hydropower, National Academy of Sciences of the Kyrgyz Republic	吉尔吉斯斯坦科学院水问题与水电研究所	【吉尔吉斯斯坦】
94	Instituto Antártico Argentino, Buenos Aires, Argentina	阿根廷南极研究所	【阿根廷】
95	International Center for Integrated Mountain Development (ICIMOD)	国际山地综合发展中心	【尼泊尔】
96	International Centre for Antarctic Information and Research	国际南极信息与研究中心	【新西兰】

续表

序号	英文名	中文名	所属国家
97	Japan Agency for Marine-Earth Science and Technology (JAMSTEC)	独立行政法人海洋研究开发机构	【日本】
98	Kazakh State National University of Al-Farabi, Almaty	哈萨克国立大学	【哈萨克斯坦】
99	Korean Polar Research Institute	韩国极地研究所	【韩国】
100	Laboratory for Ion Beam Physics, ETH	瑞士联邦理工学院离子束物理实验室	【瑞士】
101	Laboratory of Glaciology and Geophysics of the Environment (LGGE)	冰川与环境地球物理实验室	【法国】
102	Laboratory of Hydraulics, Hydrology and Glaciology, Section Glaciology, ETH	瑞士联邦理工学院水力学、水文学和冰川学部	【瑞士】
103	Main Hydrometeorological Administration of Ministry of Ecology and Extraordinary Situations of Kyrgyz Republic	吉尔吉斯斯坦生态和特别情况部水文气象局	【吉尔吉斯斯坦】
104	Max Planck Institute for Biogeochemistry	马普生物地球化学研究所	【德国】
105	Max Planck Institute for Meteorology	马普气象研究所	【德国】
106	Met Office Hadley Centre	气象局哈德来中心	【英国】
107	Mountain Research Initiative	(国际)山地研究组织	【瑞士】
108	NASA Goddard Space Flight Center: Sea Ice Remote Sensing	美国航空航天局戈达德空间飞行中心:海冰遥感	【美国】
109	National Ice Center (NIC)	国家冰中心	【美国】
110	National Ice Core Laboratory	国家冰芯实验室	【美国】
111	National Institute of Hydrology, Western Himalayan Regional Centre	西喜马拉雅区域中心国家水文研究所	【印度】
112	National Institute of Polar Research	国立极地研究所	【日本】
113	National Snow and Ice Data Center	国家雪冰数据中心	【美国】

续表

序号	英文名	中文名	所属国家
114	Niels Bohr Institute, University of Copenhagen	哥本哈根大学尼尔斯玻尔研究所	【丹麦】
115	Norwegian Meteorological Institute	挪威气象研究所	【挪威】
116	Norwegian Polar Institute	挪威极地研究所	【挪威】
117	Oeschger Centre for Climate Change Research, University of Bern	瑞士伯尔尼大学奥斯彻格气候变化研究中心	【瑞士】
118	Pakistan Institute of Engineering and Applied Sciences	巴基斯坦工程与应用科学研究所	【巴基斯坦】
119	Permafrost Institute, Russian Academy of Sciences Siberian Division	俄罗斯科学院西伯利亚分院冻土研究所	【俄罗斯】
120	Potsdam Institute for Climate Impact Research	波茨坦气候影响研究所	【德国】
121	Roald Amundsen Center for Arctic Research, University of Tromso	特鲁卢索大学罗纳多·阿蒙森北极研究中心	【挪威】
122	School of GeoSciences, University of Edinburgh	爱丁堡大学地球科学学院	【英国】
123	School of Social and Environmental Sciences, University of Dundee	邓迪大学社会与环境科学学院	【英国】
124	Scott Polar Research Institute, University of Cambridge	剑桥大学斯科特极地研究所	【英国】
125	Section Hydrology and Water Resources Management, ETH	瑞士联邦理工学院水文和水资源管理学部	【瑞士】
126	Snow and Glacier Hydrology Unit, Department of Hydrology and Meteorology	水文气象局积雪冰川水文组	【尼泊尔】
127	Snow Hydrology Group, University of California, Santa Barbara	加州大学圣塔芭芭拉分校雪水文研究组	【美国】
128	Space Applications Centre	空间应用中心	【印度】
129	Stefansson Arctic Institute, Universities in Iceland	冰岛大学斯蒂芬森北极研究所	【冰岛】

续表

序号	英文名	中文名	所属国家
130	Swiss Federal Institute of Aquatic Science and Technology	瑞士联邦水产科技研究所	【瑞士】
131	The Institute of Geography, Ministry of Education and Science, Kazakhstan	哈萨克斯坦教育与科学部地理研究所	【哈萨克斯坦】
132	The Joint Australian Centre for Astrophysical Research in Antarctica, University of New South Wales	新南威尔士大学南极洲天体物理学研究联合中心	【澳大利亚】
133	The University of Maine at Orono	缅因大学	【美国】
134	United States Geological Survey (USGS) Alaska Science Center	美国地质调查局阿拉斯加科学中心	【美国】
135	United States Army Cold Regions Research and Engineering Laboratory (CRREL)	美国陆军寒区研究和工程实验室	【美国】
136	Universidad de Magallanes (Punta Arenas), Chile	智利麦哲伦大学	【智利】
137	Universidad Federal de Río Grande do Sul, Brazil	巴西里约·格拉德苏联邦大学	【巴西】
138	Uruguayan Antarctic Institute	乌拉圭南极研究所	【乌拉圭】
139	Wadia Institute of Himalayan Geology	瓦迪亚喜马拉雅地质研究所	【印度】
140	Word Glacier Monitoring Service (WGMS)	世界冰川监测服务处	【瑞士】
141	WSL Institute for Snow and Avalanche Research SLF	瑞士积雪与雪崩研究所	【瑞士】

(2) 国内冰冻圈相关研究机构

序号	中文名	英文名
1	北京大学城市与环境学院	College of Urban and Environmental Sciences, Peking University
2	北京大学生态学系	Department of Ecology, Peking University
3	北京师范大学全球变化与地球系统科学研究院	College of Global Change and Earth System Science, Beijing Normal University
4	冰冻圈科学国家重点实验室	State Key Laboratory of Cryospheric Sciences (SKLCS)
5	成都信息工程学院高原大气与环境研究中心	Center for Plateau Atmospheric and Environmental Research, Chengdu University of Information Technology
6	大连海事大学	Dalian Maritime University
7	大连理工大学建设工程学部	Faculty of Infrastructure Engineering, Dalian University of Technology
8	冻土工程国家重点实验室	State Key Laboratory of Frozen Soil Engineering (SKLFSE)
9	国家测绘地理信息局	National Administration of Surveying, Mapping and Geoinformation
10	国家海洋局第一海洋研究所	The First Institute of Oceanography, State Oceanic Administration
11	国家气候中心	National Climate Center
12	华南师范大学地理科学学院	School of Geography, South China Normal University
13	吉林大学青藏高原地质研究中心	Research Center for Tibetan Plateau, Jilin University
14	兰州大学草地农业科技学院	School of Pastoral Agriculture Science and Technology, Lanzhou University
15	兰州大学生命科学学院	School of Life Sciences, Lanzhou University
16	兰州大学西部环境与气候变化研究院	Research School of Arid Environment & Climate Change, Lanzhou University

续表

序号	中文名	英文名
17	兰州大学资源与环境学院	College of Earth and Environment Sciences, Lanzhou University
18	南京大学地理与海洋科学学院	School of Geographic and Oceanographic Sciences, Nanjing University
19	南京师范大学	Nanjing Normal University
20	南京信息工程大学	Nanjing University of Information Science and Technology
21	西藏高原大气环境科学研究所	Tibet Institute of Plateau Atmospheric and Environmental Science
22	西藏农牧学院	Tibet Agriculture and Animal Husbandry College
23	中国地质大学(北京)青藏高原地质研究中心	Research Center for Tibetan Plateau Geology, China University of Geosciences (Beijing)
24	中国地质大学(北京)水资源与环境学院	School of Water Resources and Environment, China University of Geosciences (Beijing)
25	中国地质大学(武汉)青藏高原研究中心	Research Center for Tibetan Plateau, China University of Geosciences (Wuhan)
26	中国地质调查局	China Geological Survey
27	中国海洋大学海洋环境学院	College of Physical and Environmental Oceanography, Ocean University of China
28	中国极地研究中心	Polar Research Institute of China
29	中国科技大学地球与空间科学学院	School of Earth and Space Sciences, University of Science and Technology of China
30	中国科学院水利部成都山地灾害与环境研究所	Institute of Mountain Hazards and Environment, Chinese Academy of Sciences
31	中国科学院大气物理研究所	Institute of Atmospheric Physics, Chinese Academy of Sciences
32	中国科学院地理科学与资源研究所	Institute of Geographic Sciences and Natural Resources Research, Chinese Academy of Sciences
33	中国科学院东北地理与农业生态研究所	Northeast Institute of Geography and Agroecology, Chinese Academy of Sciences

续表

序号	中文名	英文名
34	中国科学院寒区和旱区环境和工程研究所	Cold and Arid Regions Environmental and Engineering Research Institute, Chinese Academy of Sciences
35	中国科学院南京地理与湖泊研究所	Nanjing institute of Geography and Limnology, Chinese Academy of Sciences
36	中国科学院青藏高原研究所	Institute of Tibetan Plateau Research, Chinese Academy of Sciences
37	中国科学院生态环境研究中心	Research Center for Eco-Environmental Sciences, Chinese Academy of Sciences
38	中国科学院西北高原生物研究所	Northwest Institute of Plateau Biology, Chinese Academy of Sciences
39	中国科学院新疆生态与地理研究所	XinJiang Institute of Ecology and Geography, Chinese Academy of Sciences
40	中国气象局成都高原气象研究所	Chengdu Institute of Plateau Meteorology, Chinese Meteorological Administration
41	中国气象科学研究院	Chinese Academy of Meteorological Sciences

附录5　　冰冻圈区主要科学考察站

(1)南极地区的冰冻圈研究站

序号	站名	中文名	所属国家	地理位置	海拔高度(m)
1	Aboa Station	阿博阿站	【芬兰】	73°03′S, 13°25′W	400
2	Alfred Faure, Îles Crozet	艾尔弗雷德-福尔站	【法国】	46°25′48″S, 51°51′40″E	140
3	Almirante Brown Antarctic Base	艾米兰特布朗南极基地	【阿根廷】	64°53′S, 62°53′W	10
4	Amundsen-Scott Station	阿蒙森—斯科特站	【美国】	89°59′51″S, 139°16′22″E	2830
5	Arctowski Station	阿尔茨托夫斯基站	【波兰】	62°09′34″S, 58°28′15″W	2
6	Artigas Station	阿蒂加斯站	【乌拉圭】	62°11′04″S, 58°54′09″W	17
7	Arturo Parodi Station	阿图罗—帕罗迪站	【智利】	80°19.1′S, 81°18.48′W	880
8	Asuka Station	飞鸟站	【日本】	71°31′34″S, 24°08′17″E	930
9	Belgrano Ⅱ Base	贝尔格拉诺将军2号站	【阿根廷】	77°52′29″S, 34°37′37″W	250
10	Bellingshausen Station	别林斯高晋站	【俄罗斯】	62°11′47″S, 58°57′39″W	16
11	Bernardo O'Higgins Base	伯纳德—沃伊金斯站	【智利】	63°19′15″S, 57°54′01″W	13
12	Bird Island Station	伯德岛站	【英国】	54°00′31″S, 38°03′08″W	10
13	Byrd Station	伯德站	【美国】	80°S, 119°W	1553

续表

序号	站名	中文名	所属国家	地理位置	海拔高度(m)
14	Cámara Base	卡马拉基地	【阿根廷】	62°36′S, 59°56′W	22
15	Captain Arturo Prat Base	普拉特舰长站	【智利】	62°28′45″S, 59°39′51″W	5
16	Casey Station	凯西站	【澳大利亚】	66°17′00″S, 110°31′11″E	30
17	Comandante Ferraz Station	费拉兹少校站	【巴西】	62°05′00″S, 58°23′28″W	8
18	Concordia Station	协和站	【意大利】	75°06′S, 123°21′E	3233
19	Concordia (2) Station	协和2站	【法国/意大利】	72°06′06″S, 123°23′43″E	3220
20	D10 Skiway Station	D10滑道站	【法国】	66°40′4.8″S, 139°49′10.8″E	106
21	D85 Skiway Station	D85滑道站	【法国】	70°25′30″S, 134°08′45″E	2849.88
22	Dakshin Gangotri Station	达克辛·甘戈里站	【印度】	70°05′37″S, 12°00′00″E	35
23	Dallman Station	达尔曼站	【德国】	62°14′S, 58°40′W	0
24	Davis Station	戴维斯站	【澳大利亚】	68°34′38″S, 77°58′21″E	15
25	Decepción Base	欺骗岛基地	【阿根廷】	62°58′20″S, 60°41′40″W	7
26	Dome Fuji Station	富士冰穹站	【日本】	79.00′S, 39°42.20′E	3810
27	Dumont d'Urville Station	迪蒙·迪维尔站	【法国】	66°39′46″S, 140°00′05″E	42
28	Ellsworth Station	埃尔斯沃思站	【美国/阿根廷】	77°39′S, 41°02′W	40

续表

序号	站名	中文名	所属国家	地理位置	海拔高度(m)
29	Esperanza Station	埃期佩兰站	【阿根廷】	63°23′42″S, 56°59′46″W	25
30	Estación marítima Antártica	伊斯塔肯—马丁呜站	【智利】	62°12.4′S, 58°57.45′W	5
31	Gabriel González Videla	加布里埃尔—冈萨雷斯站	【智利】	64°49.42′S, 62°51.50′W	5
32	Gondwana Station	冈瓦纳站	【德国】	74°38′S, 164°13′E	23
33	Gough Island Station	高夫岛站	【南非】	40°21′56″S, 09°52′00″W	54
34	Great Wall Station	长城站	【中国】	62°12′59″S, 58°57′44″W	10
35	Halley Station	哈雷站	【英国】	75°34′54″S, 26°32′28″W	37
36	Jinnah Station	吉纳站	【巴基斯坦】	70°24′S, 25°45′E	49
37	Juan Carlos I Station	胡安—卡洛斯 I 站	【西班牙】	62°39′S, 60°23′W	12
38	Jubany Scientific Station	尤巴尼站	【阿根廷】	62°14′16″S, 58°39′52″W	10
39	Julio Ripamonti Station	夏利奥—芮巴芒蒂站	【智利】	62°12.07′S, 58°53.13′W	50
40	Jyelcho Station	耶尔乔站	【智利】	64°62′S, 63°35′W	5
41	King Edward Point Station	爱德华王角站	【英国】	54°17′00″S, 36°29′37″W	20
42	King Sejong Station	世宗王站	【韩国】	62°13′24″S, 58°47′21″W	10
43	Kliment Ohridski Base	克雷门特—欧日德斯魁基地	【保加利亚】	62°38′29″S, 60°21′53″W	13

续表

序号	站名	中文名	所属国家	地理位置	海拔高度(m)
44	Kohnen Station	科嫩站	【德国】	75°00′S, 00°04′E	2892
45	Kunlun Station	昆仑站	【中国】	80°25′01″S, 77°06′58″W	4087
46	Luis Carvajal Station	路易斯－卡瓦哈尔站	【智利】	67°45′S, 68°54′W	10
47	Macchu Picchu Station	马丘比丘站	【秘鲁】	62°05.49′S, 58°28.27′W	10
48	Macquarie Island Station	麦夸里岛站	【澳大利亚】	54°29′58″S, 158°56′09″E	100
49	Maitri Station	麦特瑞站	【印度】	70°45′57″S, 11°44′09″E	130
50	Maldonado Base	马尔多纳多基地	【厄瓜多尔】	62°26′56.6″S, 59°44′29″W	10
51	Marambio Base	马兰比奥基地	【阿根廷】	64°14′42″S, 56°39′25″W	200
52	Mario Zucchelli Station	马里奥－组魁里站	【意大利】	74°42′S, 164°07′E	15
53	Matienzo Station	马廷索站	【阿根廷】	64°58′S, 60°03′W	32
54	Mawson Station	莫森站	【澳大利亚】	67°36′17″S, 62°52′15″E	5
55	McMurdo Station	麦克默多站	【美国】	77°50′53″S, 166°40′06″E	10
56	Melchior Station	梅尔基奥站	【阿根廷】	64°20′S, 62°59′W	8
57	Mendel Station	孟德尔站	【捷克】	63°48′16.32″S, 57°53′8.82″W	10
58	Mid Point Station	中点站	【意大利】	75°32′10″S, 145°51′32″E	2509

续表

序号	站名	中文名	所属国家	地理位置	海拔高度(m)
59	Mirny Station	和平站	【俄罗斯】	66°33′07″S, 93°00′53″E	40
60	Mizuho Station	瑞穗站	【日本】	70°41′53″S, 44°19′54″E	2230
61	Molodezhnaya Station	青年站	【俄罗斯】	67°40′18″S, 45°51′21″E	42
62	Neumayer Station Ⅲ	诺伊迈尔三号站	【德国】	70°38′00″S, 08°15′48″W	40
63	Novolazarevskaya Station	新拉扎列夫站	【俄罗斯】	70°46′26″S, 11°51′54″E	102
64	Orcadas Base	奥尔卡达斯基地	【阿根廷】	60°44′20″S, 44°44′17″W	4
65	Palmer Station	帕尔默站	【美国】	64°46′30″S, 64°03′04″W	10
66	Petrel Station	海燕站	【阿根廷】	63°28′S, 56°13′W	18
67	Presidente Frei Base	弗雷总统站	【智利】	62°12′00″S, 58°57′51″W	10
68	Primavera Station	白桃花心木站	【阿根廷】	64°09′S, 60°57′W	50
69	Princess Elisabeth Base	伊丽莎白公主基地	【比利时】	71°57′S, 23°20′E	1397
70	Professor Julio Escudero Base (Professor Escudero)	埃斯库多罗教授站	【智利】	62°12′04″S, 58°57′45″W	10
71	Progress 2 Station	进步2号站	【俄罗斯】	69°22′44″S, 76°23′13″E	64
72	Refugio Gabriel de Castilla Station	里菲吉奥加布里埃尔—卡斯蒂利亚站	【西班牙】	62°58′40″S, 60°40′30″W	15

续表

序号	站名	中文名	所属国家	地理位置	海拔高度(m)
73	Risopatrón Base	里索帕特龙基地	【智利】	62°22.92′S, 59°39.833′W	10
74	Rothera Station	罗瑟拉站	【英国】	67°34′10″S, 68°07′12″W	16
75	Russkaya Station	俄罗斯站	【俄罗斯】	74°46′S, 136°40′W	0
76	San Martin Station	圣马丁站	【阿根廷】	68°07′47″S, 67°06′12″W	5
77	Sanae Ⅳ Station	萨纳埃4号站	【南非】	71°40′25″S, 02°49′44″W	850
78	Scott Base	斯科特基地	【新西兰】	77°51′00″S, 166°45′46″E	10
79	Syowa Station	昭和站	【日本】	69°00′25″S, 39°35′01″E	29
80	Signy Station	西格尼站	【英国】	60°43′S, 45°36′W	5
81	Siple Station	赛普尔站	【美国】	75°55′S, 83°55′W	1054
82	Sitry Point Station	赛特里站	【意大利】	71°39′09″S, 148°39′20″E	2094
83	Sky Blue Station	蓝天站	【英国】	74°51′S, 71°34′W	1372
84	Sobral Station	索布拉尔站	【阿根廷】	81°05′S, 40°39′W	1000
85	Soyuz Station	联盟站	【俄罗斯】	70°35′S, 68°47′E	336
86	Svea Station	斯维站	【瑞典】	74°35′S, 11°13′W	470
87	Tor Station	托尔站	【挪威】	71°53′20″S, 05°09′30″E	1625

续表

序号	站名	中文名	所属国家	地理位置	海拔高度(m)
88	Troll (4) Station	特罗尔4站	【挪威】	72°00′07″S, 02°32′02″E	1300
89	Komcomoskaya Station	共青团站	【俄罗斯】	72°21′S, 98°00′E	3500
90	Vernadsky Station	韦尔纳茨站	【乌克兰】	65°14′43″S, 64°15′24″W	7
91	Vostok Station	东方站	【俄罗斯】	78°27′52″S, 106°50′14″E	3488
92	Wasa Station	贝森站	【瑞典】	73°03′S, 13°25′W	400
93	Zhongshan Station	中山站	【中国】	69°22′16″S, 76°23′13″E	14.9

(2)北极地区的冰冻圈研究站

序号	站名	中文名	所属国家	地理位置	海拔高度(m)
1	Abisko Naturvetenskapliga Station	阿比斯库自然科学站	【瑞典】	68°21′N, 18°49′E	385
2	Alert Station	阿勒特站	【加拿大】	82°30′N, 62°20′W	63
3	ALOMAR Station	阿罗马站	【挪威】	69°17′N, 16°01′E	380
4	Arctic Igloolik Research Center	北极冰屋研究中心	【加拿大】	69°23′N, 81°48′W	53
5	Barrow Station	巴罗站	【美国】	71°18′N, 156°46′W	3
6	Blåisen Station	艾辛河站	【挪威】	68°21′N, 17°52′E	850
7	Cainhavarre Station	肯哈瓦尔站	【挪威】	68°06′N, 18°01′E	850
8	Dasan Station	茶山站	【韩国】	78°55′N, 11°56′E	40
9	Eureka Weather Station	尤里卡天气站	【加拿大】	79°59′N, 85°57′W	10
10	Forskningsstasjonen pa Svalbard Station	斯瓦尔巴德站	【挪威】	78°50′N, 11°30′E	10
11	Gee Lake Station	季湖站	【加拿大】	69°50′N, 72°02′W	1440
12	Grise Fiord Station	格里斯峡湾站	【加拿大】	76°25′N, 82°54′W	45
13	Himadri Station	喜马椎站	【印度】	78°55′N, 11°54′E	66
14	Hornsund Station	豪恩松德站	【波兰】	77°00′N, 15°33′E	10
15	Inuvik Meteorological Station	因纽维克气象站	【加拿大】	68°22′N, 133°44′W	68

续表

序号	站名	中文名	所属国家	地理位置	海拔高度(m)
16	Iqaluit Station	努纳维特站	【加拿大】	63°45′N, 68°31′W	34
17	Kevo Subarctic Research Station	科沃亚北极站	【芬兰】	69°45′N, 27°01′E	101
18	Kilpisjärvi Biological Station	克皮斯佳维生态站	【芬兰】	69°03′N, 20°49′E	476
19	Kiruna Geophysical Observatory	可如娜地球物理观测台	【瑞典】	67°49′N, 20°20′E	452
20	Kluane Lake Research Station	可鲁湖泊站	【加拿大】	61°15′N, 138°40′W	781
21	Koldewey Station	考德威站	【德国】	78°55′24″N, 11°55′15″E	0
22	Labytnangi Ecological Research Station	拉比唐吉生态站	【俄罗斯】	66°39′N, 66°25′E	37
23	Lavrentiya Weather Station	拉维仁提亚天气站	【俄罗斯】	65°36′N, 171°03′W	39
24	McGill Arctic Research Station（M.A.R.S.）	麦吉尔北极站	【加拿大】	79°26′N, 90°46′W	176
25	Mirnoye Biological Station	莫劳叶生物站	【俄罗斯】	66°15′N, 89°00′E	77
26	Myvatn Research Station	米万站	【冰岛】	65°40′N, 17°00′W	285
27	Netherlands Arctic Station	荷兰北极站	【荷兰】	78°56′N, 11°53′E	40
28	Northeast Science Station Cherskii（RAISE）	东北切尔斯基（高地）科学站	【俄罗斯】	68°30′N, 161°32′E	16
29	Nuuk Field Station	努克站	【丹麦】	64°11′N, 51°41′W	9

续表

序号	站名	中文名	所属国家	地理位置	海拔高度(m)
30	Ny-Ålesund Arctic Research Station	新奥尔松北极站	【挪威】	78°55′N, 11°56′E	176
31	Proven-Kangerlussuaq Station/Kangerlussuaq Station	康戈鲁思瓦克站	【丹麦/格陵兰】	67°01′N, 50°41′W	108
32	Provideniya Station	皮若文登尼亚站	【俄罗斯】	64°24′N, 173°12′W	236
33	Rabben Station	洛宾站	【日本】	78°56′N, 11°52′E	40
34	Resolute Weather Station/Nunavut Arctic College	雷索柳特站	【加拿大】	74°43′N, 94°58′W	55
35	Samoylov Station	萨莫伊洛夫站	【俄罗斯/德国】	72°22′N, 126°28′E	10
36	Sermilik Station	塞尔米利克站	【丹麦】	65°41′N, 37°55′W	3216
37	Sodankylä Climate Station	苏旦可雅气候站	【芬兰】	67°22′N, 26°39′E	100
38	Station Nord	格陵兰北方站	【丹麦】	81°36′N, 16°40′W	76
39	Storsteinsfiell Station	大石山站	【挪威】	68°13′N, 17°55′E	975
40	Summit Camp	顶点营地	【美国】	72°34′46.5″N, 38°27′33.07″W	3216
41	Tåkeheim (Engabreen) Station	恩噶伯仁站	【挪威】	66°40′N, 13°51′E	980
42	Tarfala Station	塔法拉站	【瑞典】	67°55′N, 18°40′E	950
43	Teriberka Station	捷里别尔卡站	【俄罗斯】	69°1.2′N, 35°7.2′W	40

续表

序号	站名	中文名	所属国家	地理位置	海拔高度(m)
44	Thule Station	图里站	【丹麦】【美国】【意大利】	76°24′N, 68°43′W	786
45	Toolik Field Station	图利克站	【美国】	68°38′N, 149°36′W	720
46	Tuktoyaktuk Field Station	图克妥亚库克站	【加拿大】	69°26′N, 133°02′W	0
47	Tundra Ecological Research Station	苔原生态站	【加拿大】	64°52′N, 111°34′W	424
48	University of Buffalo Brooks Range Camp	布法罗大学布鲁克斯山脉站	【美国】	68°06′N, 149°27′W	1130
49	Versailles (Trollbergdalsbreen) Station	凡尔赛站	【挪威】	66°43′N, 14°27′E	619
50	Whapmagoostui-Kuujjuarapik Research Station	瓦泊玛古斯图—库娟如皮克站	【加拿大】	55°17′N, 77°45′W	32
51	White Sea Biological Station	白海生物站	【俄罗斯】	66°34′N, 33°08′E	45
52	Willem Barents Field Station	威勒巴伦支站	【俄罗斯】	73°22′N, 80°32′E	0
53	Wrangel Island Station	旺格岛站	【俄罗斯】	71°20′N, 179°30′E	137
54	Yellow River Station	黄河站	【中国】	78°55′N, 11°56′E	62
55	Zackenberg Ecological Research Station	柴肯堡生态站	【格陵兰】	74°28′11″N, 0°34′25″W	30

(3)其他地区的冰冻圈研究站

序号	站名	中文名	所属国家	地理位置	海拔高度(m)
1	Ålfoten Station	奥勒佛腾站	【挪威】	61°45′N, 5°40′E	905
2	Alliance Station	联盟站	【肯尼亚】	0°N, 37°E	4000
3	Alpine Ecosystem Observation and Experiment Station of Mt. Gongga	贡嘎山高山生态系统观测试验站	【中国】	29°36′N, 101°53′E	3600
4	Atmosphere Watch Baseline Observatory Station in Mt. Waliguan	青海瓦里关国家大气成分本底观测研究站	【中国】	36°17′N, 100°54′E	3816
5	Beiluhe Observation and Research Station on Frozen Soil Engineering and Environment in Qinghai-Tibet Plateau	青藏高原北麓河冻土工程与环境综合观测研究站	【中国】	34°51.236′N, 92°56.395′E	4628
6	Bomi Geological Hazards Observation and Research Station	波密地质灾害观测研究站	【中国】	29°52′N, 95°46′E	2720
7	Carrick Station	大帆船站	【新西兰】	45°S, 169°E	1300
8	Cryosphere Research Station on Qinghai-Xizang Plateau	中国科学院青藏高原冰冻圈观测研究站(国家站)	【中国】	36°23′37″N, 94°54′29″E	2824
9	Dart Glacier Station	达特冰川站	【新西兰】	44°30′S, 168°37′E	1425
10	Dongchuan Debris Flow Observation and Research Station, CAS	中国科学院东川泥石流观测研究站	【中国】	26°14′N, 103°08′E	1320

续表

序号	站名	中文名	所属国家	地理位置	海拔高度(m)
11	Dzhankuat Glacier Station	占库纳特冰川站	【美国】	43°12′N, 43°46′E	2700
12	Eidgenössischen Institutes für Schnee-und Lawinenforschung	联邦研究所雪和雪崩研究站	【瑞士】	46°50′N, 9°48′E	2667
13	Engabreen Station	昂噶冰川站	【挪威】	61°49′N, 7°16′E	900
14	Gråsubreen Station	格拉苏冰川站	【挪威】	61°39′N, 8°36′E	2100
15	Gulkana Glacier Station	库尔卡纳冰川站	【美国】	63°16′N, 145°25′W	1800
16	Heihe Remote Sensing Observation Station	黑河遥感试验站	【中国】	38°10′10″N, 100°28′40″E	1524
17	Hidden Valley Station	隐谷站	【尼泊尔】	28°47.5′N, 83°33′E	5055
18	Hintereis Station	海因特冰川站	【奥地利】	46°47′47″N, 10°45′48″E	3026
19	Icefield Range Research Station	冰原山脉站	【加拿大】	61°01′N, 138°24′W	770
20	Institute of Arctic and Alpine Research (INSTAAR) Mountain Research Station	北极与高山研究站	【美国】	40°02′13″N, 105°32′39″W	2925
21	Lhajung Station	拉樟站	【尼泊尔】	27°53.8′N, 86°49.6′E	4420
22	Mascardi-Tronador Station	马斯克蒂—绰纳德研究站	【阿根廷】	41°16′S, 71°38′W	845
23	Moshiri Station	茂尻站	【日本】	43°20′N, 142°05′E	290

续表

序号	站名	中文名	所属国家	地理位置	海拔高度(m)
24	Muztagh Ata Station for Westerly Environment Observation and Research	慕士塔格西风带环境综合观测研究站	【中国】	38°24′N, 75°02′E	3660
25	Nam Co Monitoring and Research Station for Multisphere Interactions	纳木错圈层相互作用综合观测研究站	【中国】	30°46′N, 90°59′E	4730
26	Nagqu Station of Plateau Climate and Environment (NPCE)	那曲高寒气候环境观测研究站	【中国】	31.37°N, 91.90°E	4509
27	Ngari Station for Desert Environment Observation and Research	阿里荒漠环境综合观测研究站	【中国】	33°24′N, 79°42′E	4264
28	Nigardsbreen Station	尼加斯冰川站	【挪威】	61°43′N, 7°08′E	1350
29	Niwot Ride Long-Term Ecological Research Site	妮沃特山脉长期生态站	【美国】	40°02′56″N, 105°38′24″W	3814
30	Okstindsjøen Field Station	奥克斯汀由恩站	【挪威】	62°02′N, 14°25′E	760
31	Optimus (Vesledalsbreen) Station	擎天柱站	【挪威】	61°51′N, 7°16′E	1573
32	Peyton Glacier Station	佩顿冰川站	【加拿大】	51°40′N, 116°35′W	2220
33	Qilian Alpine Ecology & Hydrology Research Station	黑河上游生态—水文试验研究站	【中国】	38°15′N, 99°52′E	3040
34	Qilian Shan Station of Glaciology and Ecologic Environment	祁连山冰川与生态环境观测研究站	【中国】	39°30.6′N, 96°30.6′E	4180

续表

序号	站名	中文名	所属国家	地理位置	海拔高度(m)
35	Qomolangma Atmospheric and Environmental Observation and Research Station	珠穆朗玛峰大气与环境综合观测研究站	【中国】	28°12.6′N, 86°33.6′E	4276
36	Quelccaya Station	奎珂雅站	【秘鲁】	13°56′S, 70°50′W	5645
37	Ravnarheim (φstre Memurûbre) Station	拉乌那艾姆(东麦穆鲁尔)站	【挪威】	61°33′N, 8°31′E	1870
38	Sonnblick Observatory	松布利克山观象台	【奥地利】	47°01′N, 13°12′E	3105
39	Southeast Tibet Observation and Research Station for the Alpine Environment	藏东南高山环境综合观测研究站	【中国】	29°46′N, 01°44′E	3230
40	Sperry Chalet Field Project	斯佩里站	【美国】	48°37′N, 113°52′W	2469
41	Tianshan Glaciological Station	天山冰川国家野外科学观测研究站	【中国】	43°03′N, 86°29.4′E	2130
42	Tianshan Station for Snow and Avalanche Research	天山积雪雪崩研究站	【中国】	43°16′N, 84°24′E	1776
43	Toikanbetsu Station	托坎别次站	【日本】	44°5′44″N, 142°1′32″E	235
44	Tomakomai Station	托马阔买站	【日本】	42°27′N, 141°45′E	15
45	Tordenfjell (Hellstugubreen) Station	托登山(埃尔斯托古布林)站	【挪威】	61°34′N, 8°27′E	2085
46	Tsurgisawa-Goya Station	茨鲁圪萨瓦高亚站	【日本】	36°36′N, 137°37′E	2460

续表

序号	站名	中文名	所属国家	地理位置	海拔高度(m)
47	Variegated Glacier Station	花冰川站	【美国】	59°59′54″N, 139°19′54″W	1200
48	Wolverine Glacier Station	沃尔弗林冰川站	【美国】	60°23′N, 148°55′W	900
49	Wrangell Station	兰格尔站	【美国】	62°N, 144°W	4000
50	Xainza Alpine Steppe and Wetland Ecosystem Observation Station	申扎高寒草原与湿地生态系统观测试验站	【中国】	30°57′N, 88°42′E	4675
51	Yellow River Water Resource Region Climate and Environment Comprehesive Observation and Research Station	玛曲黄河源区气候与环境综合观测研究站	【中国】	33°51′N, 102°12′E	3430
52	Yulong Snow Mountain Glacial and Environmental Observation Station	玉龙雪山冰川与环境观测研究站	【中国】	27°10′N, 100°13′E	2400

附录6 主要学术组织(冰冻圈相关)

序号	英文	中文
1	American Geophysical Union (AGU)	美国地球物理联合会
2	American Meteorological Society (AMS)	美国气象学会
3	Association of Polar Early Career Scientists (APECS)	极地青年科学家协会
4	Chinese Association for Quaternary Research	中国第四纪科学研究会
5	Chinese Geophysical Society (CGS)	中国地球物理学会
6	Chinese Hydraulic Engineering Society (CHES)	中国水利学会
7	Chinese Meteorological Society (CMS)	中国气象学会
8	European Geosciences Union (EGU)	欧洲地球科学协会
9	Geological Society of China	中国地质学会
10	Intergovernmental Oceanographic Commission (IOC)	政府间海洋学委员会
11	International Antarctic Meteorological Research Committee (IAMRC)	国际南极气象研究委员会
12	International Arctic Science Committee (IASC)	国际北极科学委员会
13	International Association of Cryospheric Sciences (IACS)	国际冰冻圈科学协会
14	International Association of Hydrological Sciences (IAHS)	国际水文科学协会
15	International Association of Meteorology and Atmospheric Sciences (IAMAS)	国际气象和大气科学协会
16	International Commission on Polar Meteorology (ICPM)	国际极地气象学委员会
17	International Geographical Union (IGU)	国际地理联合会
18	International Glaciological Society (IGS)	国际冰川学会
19	International Permafrost Association (IPA)	国际冻土学协会

续表

序号	英文	中文
20	Japanese Society of Snow and Ice (JSSI)	日本雪冰学会
21	Royal Meteorological Society (RMetS)	英国皇家气象学会
22	Scientific Committee on Antarctic Research (SCAR)	南极研究科学委员会
23	The China Society on Tibet Plateau (CSTP)	中国青藏高原研究会
24	The Geographical Society of China (GSC)	中国地理学会

附录7 主要科学计划(冰冻圈相关)

序号	英文	中文
1	Antarctic Mapping Mission (AMM)	南极制图计划
2	Arctic Circumpolar Coastal Observatory Network (ACCOnet)	环北极海岸监测计划
3	Arctic Climate System Study (ACSYS)	北极气候系统研究计划
4	Arctic Ice Dynamic Experiment (AIDEX)	北极冰层动力学试验
5	Arctic Ice Dynamics Joint Experiment (AIDJEX)	北极冰动力学联合试验
6	Arctic Monitoring and Assessment Programme (AMAP)	北极监测和评估计划
7	Atmospheric Model Intercomparison Project (AMIP)	大气模式比较计划
8	Atmospheric Radiation Measurement Program (ARM)	大气辐射测量计划
9	Bedrock Mapping Project (BEDMAP)	南极冰下基岩地形制图计划
10	Boreal Ecosystem-Atmosphere Study (BOREAS)	北方生态-大气研究计划
11	Carbon Pools in Permafrost Regions Project (CAPP)	冻土地区碳汇研究计划
12	Circumpolar Active Layer Monitoring (CALM)	环北极活动层监测计划
13	Climate and Cryosphere (CliC)	气候与冰冻圈计划
14	Climate Variability and Predictability (CLIVAR)	气候变化与可预报性研究计划
15	Cooperative Holocene Mapping Project (COMAP)	全新世综合制图计划
16	Coupled Model Intercomparison Project (CMIP)	耦合模式比较计划
17	Cryospheric Network (CryoNet)	冰冻圈网络
18	Defense Meteorological Satellite Program (DMSP)	国防气象卫星计划

续表

序号	英文	中文
19	Developing Arctic Modelling and Observing Capabilities for Long-term Environmental Studies (DAMOCLES)	发展的北极模型和观测能力的长期环境研究计划
20	DIVERSITAS	国际生物多样性计划
21	Eco-hydrological Observational Experiments of the Heihe River Basin (EOEHR)	黑河流域生态－水文观测试验
22	European Global Precipitation Mission (EGPM)	欧洲全球降水计划
23	Global Change System for Analysis Research and Training (START)	全球变化分析研究和培训计划
24	Global Climate Observing System (GCOS)	全球气候观测系统
25	Global Cryosphere Watch (GCW)	全球冰冻圈监测
26	Global Earth Observation System of Systems (GEOSS)	全球综合地球观测系统
27	Global Energy and Water Cycle Experiment (GEWEX)	全球能量与水循环试验
28	Global Environment Monitoring Program (GEMS)	全球环境监测计划
29	Global Environmental Change (GEC)	全球环境变化计划
30	Global Land Ice Measurements from Space (GLIMS)	全球陆地冰空间监测计划
31	Global Monitoring for Environment and Security (GMES)	全球环境和安全监测计划
32	Global Observation System (GOS)	全球观测系统
33	Global Ocean Observing System (GOOS)	全球海洋观测系统
34	Global Terrestrial Network for Glaciers (GTN-G)	全球陆地监测网－冰川
35	Global Terrestrial Network for Permafrost (GTN-P)	全球陆地监测网－冻土
36	Global Water System Project (GWSP)	全球水系统计划
37	Greenland Ice-core Project (GRIP)	格陵兰冰芯计划

续表

序号	英文	中文
38	Greenland Ice Sheet Project 2 (GISP2)	格陵兰冰芯计划 II
39	Indian Ocean Experiment (INDOEX)	印度洋试验
40	Integrated Arctic Ocean Observing System (IAOOS)	北冰洋综合观测系统
41	International Arctic Buoy Programme (IABP)	国际北极浮标计划
42	International Geosphere-Biosphere Programme (IGBP)	国际地圈生物圈计划
43	International Global Water Cycle Observation (IGWCO)	国际全球水循环观测
44	International Human Dimensions Programme (IHDP)	国际人文因素计划
45	International Polar Year (IPY)	国际极地年
46	International Programme for Antarctic Buoy (IPAB)	国际南极浮标计划
47	International Satellite Cloud Climatology Project (ISCCP)	国际卫星云气候计划
48	International Trans-Antarctic Scientific Expedition (ITASE)	国际横穿南极科学计划
49	International Tundra Experiment (ITEX)	国际苔原试验计划
50	Long-term Ecological Research (LTER)	长期生态研究计划
51	Model Intercomparison Project (MIP)	模型比对计划
52	North Greenland Eemian Ice Drilling (NEEM)	格陵兰北部末次间冰期冰芯钻取计划
53	Past Global Changes (PAGES)	过去全球变化研究计划
54	Permafrost Young Researchers Network (PYRN)	多年冻土学青年研究者网络计划
55	Program for Climate Model Diagnosis and Intercomparison (PCMDI)	气候模式诊断和比较计划
56	Scientific Survey and Research of the Qinghai-Tibet Plateau (SSRQTP)	青藏高原科学考察计划

续表

序号	英文	中文
57	Snow Model Intercomparison Project (SnowMIP)	雪模式对比计划
58	Third Pole Environment (TPE)	"第三极环境"科学计划
59	US Study of Environmental Arctic Change (SEARCH)	美国北极环境变化研究计划
60	WMO Integrated Global Observing System (WIGOS)	世界气象组织集成全球观测系统
61	World Climate Research Programme (WCRP)	世界气候研究计划
62	World Hydrological Cycle Observing System (WHYCOS)	世界水文循环观测系统

附录8　与冰冻圈相关的主要缩略词

序号	缩略语	英文	中文
1	AABW	Antarctic Bottom Water	南极底层水
2	AAD	artificial additional dose	人工附加剂量
3	AAQS	Ambient Air Quality Standard	环境空气质量标准
4	AAR	accumulation area ratio	积累区面积比率
5	AAR	Amino Acid Racemization	氨基酸外消旋法
6	AATSR	Advanced Along-Track Scanning Radiometer	高级沿轨扫描辐射计
7	ABC	Atmospheric Brown Cloud	大气棕色云
8	ABL	atmospheric boundary layer	大气边界层
9	ACC	Antarctic Circumpolar Current	南极绕极流
10	ACCOnet	Arctic Circumpolar Coastal Observatory Network	环北极海岸监测计划
11	ACR	Antarctic Cold Reversal	南极降温逆转
12	ACSYS	Arctic Climate System Study	北极气候系统研究计划
13	A/D	analog-to-digital conversion	模—数转换器
14	AD	archaeological dose	考古剂量
15	AEROS	Advanced Earth Resources Observation System	高级地球资源观测系统
16	AH	Azores High	亚速尔高压
17	AIDEX	Arctic Ice Dynamic Experiment	北极冰层动力学试验
18	AIDJEX	Arctic Ice Dynamics Joint Experiment	北极冰动力学联合试验

续表

序号	缩略语	英文	中文
19	AIMR	Airborne Imaging Microwave Radiometer	机载成像微波辐射计
20	AINS	aircraft inertial navigation system	飞机惯性导航系统
21	ALOS	Advanced Land Observation System	先进陆地观测卫星
22	AMAP	Arctic Monitoring and Assessment Programme	北极监测和评估计划
23	AMIP	Atmospheric Model Intercomparison Project	大气模式比较计划
24	AMM	Antarctic Mapping Mission	南极制图计划
25	AMS	accelerator mass spectrometry	加速器质谱测量
26	AMSR	Advanced Microwave Scanning Radiometer	高级微波扫描辐射计
27	AMSR-E	Advanced Microwave Scanning Radiometer for Earth Observing System	地球观测系统高级微波扫描辐射计
28	AMSU	Advanced Microwave Sounding Unit	先进微波探测器
29	ANS	aircraft navigation system	飞机导航系统
30	AO	Arctic Oscillation	北极涛动
31	APF	Antarctic Polar Front	南极极锋
32	Aqua	Aqua satellite	美国国家航空航天局水卫星
33	ARMP	Atmospheric Radiation Measurement Program	大气辐射测量计划
34	ASAR	Advanced Synthetic Aperture Radar	先进合成孔径雷达
35	ASCAS	Alpine Snow-Cover Analysis System	高山积雪分析系统
36	ASF	Alaska SAR Facility	阿拉斯加雷达地面站
37	ASTER	Advanced Spaceborne Thermal Emission and Reflection Radiometer	先进星载热发射反射辐射计

续表

序号	缩略语	英文	中文
38	ATCPs	Antarctic Treaty Consultative Parties	南极条约协商国
39	ATLAS	Advanced Topographic Laser Altimeter System	先进地形激光测高计系统
40	ATS	application technology satellite	应用技术卫星
41	ATSR	Along Track Scanning Radiometer	沿轨扫描辐射计
42	AVA	anticyclone vorticity advection	反气旋涡度平流
43	AVHRR	Advanced Very High Resolution Radiometer	高级甚高分辨率辐射计
44	AVIRIS	Airborne Visible/Infrared Imaging Spectrometer	机载可见光/红外成像分光计
45	AVNIR-2	Advanced Visible and Near Infrared Radiometer type 2	高级可见与近红外辐射计二型
46	AWS	automatic weather station	自动气象站
47	BAW	Bolling-Allerod Warming Period	波林—阿洛德暖期
48	B.C.	Before Christ	公元前
49	BEDMAP	Bedrock Mapping Project	南极冰下基岩地形制图计划
50	BOD	biochemical oxygen demand	生化需氧量
51	BOREAS	Boreal Ecosystem-Atmosphere Study	北方生态—大气研究计划
52	B.P.	Before Present	距今(今,指公元1950年)
53	BRDF	bidrectional reflectance distribution function	双向反射率分布函数
54	CAAA	Clean Air Act Amendments	空气清洁法修正案
55	CALM	Circumpolar Active Layer Monitoring	环北极活动层监测计划
56	CAP	Common Agricultural Policy	(欧共体)共同农业政策

续表

序号	缩略语	英文	中文
57	CAPP	Carbon Pools in Permafrost Regions Project	冻土地区碳汇研究计划
58	CAT	clear air turbulence	晴空湍流
59	CBA	cost-benefit analysis	成本－效率分析
60	CBERS	China-Brazil Earth Resources Satellite	中巴地球资源卫星
61	CCD	charge-coupled device	电荷耦合器件
62	CCL	convective condensation level	对流凝结高度
63	CCN	cloud condensation nuclei	云凝结核
64	CDE	carbon dioxide equivalence	二氧化碳当量
65	CDM	Clean Development Mechanism	清洁发展机制
66	CEOP	Cooperated Enhanced Observing Period	全球协调加强观测期
67	CEOS	Committee on Earth Observation Satellites	地球观测卫星委员会
68	CFCs	chlorofluorocarbons	氯氟碳化合物
69	CFM	chlorofluoromethane	氟利昂
70	CGSC	Chinese Geodetic Stars Catalogue	中国大地测量星表
71	CGT	China Grassland Transect	中国草地样带
72	CHRIS	Compact High Resolution Imaging Spectrometer	紧密型高分辨率成像光谱仪
73	CIO	Conventional International Origin	国际协议原点
74	CL	condensation level	凝结高度
75	CliC	Climate and Cryosphere	气候与冰冻圈计划
76	CLIVAR	Climate Variability and Predictability	气候变化与可预报性研究计划

续表

序号	缩略语	英文	中文
77	CLPX	Cold Land Processes Experiment	寒区陆面过程野外试验
78	CMBR	cosmic microwave background radiation	宇宙微波背景辐射
79	CME	coronal mass ejection	日冕物质抛射
80	CMIP	Coupled Model Intercomparison Project	耦合模式比较计划
81	CMP	common-midpoint	共中心点
82	COD	chemical oxygen demand	化学需氧量
83	COF	crystal fabric orientation	冰晶（主轴）取向
84	COMAP	Cooperative Holocene Mapping Project	全新世综合制图计划
85	COSMO-SkyMed	Constellation of Small satellites for Mediterranean Basin Observation	地中海区域小卫星群观测
86	CPEC	close-path eddy covariance	闭路涡度相关
87	CryoNet	Cryospheric Network	冰冻圈网络
88	CSCD	continuous snow cover duration	连续积雪期
89	D/A	digital-to-analog conversion	数—模转换
90	DAMOCLES	Developing Arctic Modelling and Observing Capabilities for Long-term Environmental Studies	发展的北极模型和观测能力的长期环境研究计划
91	DAS	data acquisition system	数据采集系统
92	DBF	deciduous broadleaf forest	落叶阔叶林
93	DDM	degree day model	度日模型
94	DEM	digital elevation model	数字高程模型
95	DEP	dielectric profiling	介电特征剖面
96	DIC	dissolved inorganic carbon	溶解态无机碳

续表

序号	缩略语	英文	中文
97	DISORT	discrete ordinate radiative transfer model	离散坐标辐射传输模型
98	DMS	dimethyl sulfide	二甲基硫
99	DMSO	dimethyl sulfoxide	二甲基硫亚砜
100	DMSP	Defense Meteorological Satellite Program	国防气象卫星计划
101	DNF	deciduous needleleaf forest	落叶针叶林
102	DOC	dissolved organic carbon	溶解态有机碳
103	D-O cycle	Dansgaard-Oeschger cycle D-O cycle	D-O 旋迴
104	D-O event	Dansgaard-Oeschger event D-O event	D-O 事件
105	DOM	digital orthophoto map	数字正射影像图
106	D-O Oscillations	Dansgaard-Oeschegar Oscillations	D-O 波动
107	DORIS	Doppler Orbitograph and Radio Positioning Intergrated by Satellite	多里斯系统
108	DSM	digital surface model	数字表面模型
109	DSS	decision support system	决策支持系统
110	DTP	data transport protocol	数据传输协议
111	EACGT	Euro-Asian Continental Grassland Transect	欧亚大陆草地样带
112	EAIS	East Antarctic Ice Sheet	东南极冰盖
113	EBF	evergreen broadleaf forest	常绿阔叶林
114	ECD	electrochemical detector	电化学检测器
115	EC	eddy covariance	涡度相关法

续表

序号	缩略语	英文	中文
116	EC	electrical conductivity	电导性
117	EFZ	echo-free zone	无回波带,回波空白带
118	EGPM	European Global Precipitation Mission	欧洲全球降水计划
119	ELA	equilibrium line altitude	平衡线高度
120	ENF	evergreen needleleaf forest	常绿针叶林
121	ENSO	El Niño/Southern Oscillation	厄尔尼诺/南方涛动
122	ENVISAT	ENVIronmental SATllite	环境卫星
123	EOEHR	Eco-hydrological Observational Experiments of the Heihe River Basin	黑河流域生态—水文观测试验
124	EOF	empirical orthogonal function	经验正交函数
125	EOS	Earth Observing System	地球观测系统
126	ERBS	Earth Radiation Budget Satellite	地球辐射收支测量卫星
127	ERS	Earth Resource Satellite	中国地球资源卫星
128	ERS	Environmental Research Satellite	美国环境研究卫星
129	ERS	European Remote-sensing Satellite	欧洲遥感卫星
130	ERT	electrical resistivity tomography	电阻率成像[技术]
131	ERTS	Earth Resource Technology Satellite	地球资源技术卫星
132	ERU	emission reduction unit	排放减量单位
133	ESA	environmental sensitive area	环境敏感区
134	ESA	European Space Agency	欧洲空间局
135	ESMR	electrically scanned microwave radiometer	电子扫描微波辐射计

续表

序号	缩略语	英文	中文
136	ESR	electron spin resonance	电子自旋共振
137	ETM+	Enhanced Thematic Mapper Plus	增强型专题制图仪
138	FACE	Free-Air Carbon Dioxide Enrichment	自由大气二氧化碳富集
139	FFT	fast Fourier transform	快速傅里叶变换
140	FMCW	frequency-modulated continuous-wave	调频连续波
141	FT dating	fission track dating	裂变径迹测年
142	GAW	Global Atmosphere Watch	全球大气监测网
143	GCM	general circulation model	大气环流模式
144	GC-MS	gas chromatography-mass spectrometer	气相色谱－质谱仪
145	GCOM	Global Change Observation Mission	全球变化观测计划
146	GCOS	Global Climate Observing System	全球气候观测系统
147	GCP	Global Carbon Project	全球碳计划
148	GCP	ground control point	地面控制点
149	GCTE	Global Change and Terrestrial Ecosystems	全球变化与陆地生态系统
150	GCW	Global Cryosphere Watch	全球冰冻圈监测
151	GEC	Global Environmental Change	全球环境变化计划
152	GEMS	Global Environment Monitoring Program	全球环境监测计划
153	GEOS	Geodetic and Earth Orbiting Satellite	大地测量地球轨道卫星
154	GEOS	Geodynamics Experimental Ocean Satellite	地球动力学实验海洋卫星
155	GEOSS	Global Earth Observation System of Systems	全球对地观测系统之系统

续表

序号	缩略语	英文	中文
156	GEWEX	Global Energy and Water Cycle Experiment	全球能量与水循环实验
157	GHGs	greenhouse gases	温室气体
158	GIS	geographic information systerm	地理信息系统
159	GISP2	Greenland Ice Sheet Project 2	格陵兰冰芯计划 II
160	GLAS	geoscience laser altimeter system	地球科学激光测高系统
161	GLCM	Gray-Level Co-occurrence Matrix	灰度共生矩阵
162	GLDAS	Global Land Data Assimilation System	全球陆面数据同化系统
163	GLIMS	Global Land Ice Measurements from Space	全球陆地冰空间监测计划
164	GLOF	glacier lake outburst flood	冰湖突发洪水
165	GLRS	Geoscience Laser Ranging System	参见 GLAS(地球科学激光测高系统)
166	GMES	Global Monitoring for Environment and Security	全球环境和安全监测计划
167	GMS	Geostationary Meteorological Satellite	日本地球同步气象卫星
168	GNSS	Global Navigation Satellite System	全球导航卫星系统
169	GOES	Geostationary Operational Environmental Satellite	美国地球静止业务环境卫星
170	GOMS	Geostationary Operational Meteorological Satellite	美国地球静止业务气象卫星
171	GOOS	Global Ocean Observing System	全球海洋观测系统
172	GOS	Global Observation System	全球观测系统
173	GPP	gross primary productivity	总初级生产力
174	GPR	ground penetrating radar	探地雷达

续表

序号	缩略语	英文	中文
175	GPS	Global Positioning System	全球定位系统
176	GRIP	Greenland Ice-core Project	格陵兰冰芯计划
177	GSM	Gamburtsev subglacial mountains	甘伯采夫冰下山脉
178	GTC	gecoded terrain corrected	地形校正编码
179	GTN-G	Global Terrestrial Network for Glaciers	全球陆地监测网—冰川
180	GTN-P	Global Terrestrial Network for Permafrost	全球陆地监测网—冻土
181	GTS	gloal telecommunication system	全球电信系统(气象)
182	GWP	global warming potential	全球增温潜能
183	GWSP	Global Water System Project	全球水系统计划
184	HAWs	high-altitude wetlands	高海拔湿地
185	HCFCs	hydrochlorofluorocarbons	氢氯氟碳化物
186	HCMM	Heat Capacity Mapping Mission	热容量成像卫星
187	HIRIS	HIgh Resolution Imaging Spectrometer	高分辨率成像光谱仪
188	HIS	Hyperspectral Imaging Sensor	超光谱成像仪
189	HISS	Holographic Ice Surveying System	全息冰层探测系统
190	HKH	Hindu Kush-Karakorum-Himalaya	兴都库什—喀喇昆仑—喜马拉雅
191	HPI	human poverty index	人类贫困指数
192	IABP	International Arctic Buoy Programme	国际北极浮标计划
193	IAHS	International Association of Hydrologic Sciences	国际水文科学协会
194	IAOOS	Integrated Arctic Ocean Observing System	北冰洋综合观测系统

续表

序号	缩略语	英文	中文
195	IAT	International Atomic Time	国际原子时
196	ICESat	Ice, Cloud, and land Elevation Satellite	冰、云和陆地高程卫星
197	ICRF	International Celestial Reference Frame	国际天球参考系
198	IFOV	instantaneous field of view	瞬时视场
199	IGBP	International Geosphere-Biosphere Programme	国际地圈生物圈计划
200	IGWCO	Integrated Global Water Cycle Observation	全球综合水循环观测
201	IGY	International Geophysical Year	国际地球物理年
202	IHDP	International Human Dimensions Programme	国际人文因素计划
203	IHO	International Hydrography Organization	国际航道测量组织
204	IIP	International Ice Patrol	国际冰情巡逻队
205	INDOEX	Indian Ocean Experiment	印度洋试验
206	InSAR	Interferometric Synthetic Aperture Radar	干涉合成孔径雷达
207	INS	Inertial Navigation System	惯性导航系统
208	IPAB	International Programme for Antarctic Buoy	国际南极浮标计划
209	IPM	integrated pest management	有害生物综合治理
210	IPY	International Polar Year	国际极地年
211	IRH	internal reflection horizon	内部反射层
212	IRSDU	ice resistant semisubmersible drilling unit	抗冰半潜式平台/装置
213	IRS	Indian Remote Sensing Satellite	印度遥感卫星

续表

序号	缩略语	英文	中文
214	IRS	Infrared Spectrograph	红外多光谱成像仪
215	ISCCP	International Satellite Cloud Climatology Project	国际卫星云气候计划
216	ISDN	integrated services digital network	数字通信网络
217	ISS	inertial surveying system	惯性测量系统
218	ITASE	International Trans-Antarctic Scientific Expedition	国际横穿南极科学计划
219	ITCZ	Inter-Tropical Convergence Zone	热带辐合带
220	ITEX	International Tundra Experiment	国际苔原试验计划
221	ITOS	Improved TIROS Operational Satellite	改进型"泰罗斯"业务卫星
222	ITRF	International Terrestrial Reference Frame	国际地球参考框架
223	JAWS	Joint Arctic Weather Stations	北极天气联合站网
224	JERS-1	Janpanese Earth Resources Satellite 1	日本地球资源卫星1号
225	JPL	Jet Propulsion Laboratory	喷气推进实验室
226	K/Ar dating	Patassium-Argon dating	钾氩测年
227	KP	Kyoto Protocol	京都议定书
228	LAI	leaf area index	叶面积指数
229	LAM	limited area model	有限区域模式
230	Landsat	Land Satellite	美国陆地卫星
231	LAS	Large Aperture Scintillometer	大孔径闪烁仪
232	LCCS	Land Cover Classification System	土地覆盖分类系统
233	LCL	lifting condensation level	抬升凝结高度

续表

序号	缩略语	英文	中文
234	LFC	large format camera	大像幅摄影机
235	LGM	last glacial maximum	末次冰期冰盛期
236	LIA	Little Ice Age	小冰期
237	LiDAR	Light Detection and Ranging	激光雷达
238	LLR	lunar laser ranging	激光测月
239	LOS	Land Observation Satellite	陆地观测卫星
240	LTER	Long-term Ecological Research	长期生态研究计划
241	MARPOL	International Convention for the Prevention of Pollution from Ships	防止船舶污染国际公约
242	MCP	Medieval Climate Optimum	中世纪气候适宜期
243	MERIS	MEdium Resolution Imaging Spectrometer	中分辨率光谱仪
244	METEOSAT	Meteorological Satellite	法国气象卫星
245	MIP	Model Intercomparison Project	模型比对计划
246	MIS	marine isotope stage	海洋同位素阶段
247	MISR	Multi-angle Imaging SpectroRadiometer	多角度成像光谱仪
248	MLA	Multilateral Recognition Arrangement	多边协议
249	MMA	mean annual accumulation	年平均积累率
250	MOC	meridional overturning circulation	经向反转环流
251	MODIS	Moderate Resolution Imaging Spectroradiometer	中分辨率成像光谱仪
252	MOIS	Marine Oxygen Isotope Stage	海洋氧同位素阶段
253	MOS	Marine Observation Satellite	海洋观测卫星

续表

序号	缩略语	英文	中文
254	MPT	Mid-Pleistocene transition	中更新世气候转型
255	MSA	methanesulfonic acid	甲基磺酸
256	MSAR	Multifrequency SAR	多频率合成孔径雷达
257	MSSCC	Multicolor Spin-Scan Cloud Cover Camera	多色自旋扫描云覆盖相机
258	MSS	multispectral scanner	多波段扫描仪
259	MSU	Microwave Sounding Unit	微波探测器
260	MSY	maximum sustainable yield	最大可持续产量
261	MTB	mountain bike	山地自行车
262	MTF	modulation transfer function	调制传递函数
263	MTSAT	Multi-functional Transport Satellite	（日本）多功能运输卫星
264	MWHS	Microwave Humidity Sounder	（风云三号卫星的）微波湿度辐射计
265	MWP	Medieval Warm Period	中世纪暖期
266	MWRI	Microwave Radiometer Imager	（风云三号卫星的）微波成像仪
267	NAO	North Atlantic Oscillation	北大西洋涛动
268	NAT	nitric acid trihydrate	硝酸三水合物
269	NBL	nocturnal boundary layer	夜间边界层
270	NBP	net biome production	净生物群系生产量
271	NBP	net biome productivity	净生物群系生产力
272	NDCI	Normalized Differential Cloud Index	归一化云指数
273	NDSII	Normalized Difference Snow and Ice Index	归一化冰雪指数

续表

序号	缩略语	英文	中文
274	NDSI	Normalized Difference Snow Index	归一化积雪指数
275	NDVI	Normalized Difference Vegetation Index	归一化植被指数
276	NDWI	Normalized Difference Water Index	归一化水体指数
277	NECT	NorthEastern China Transect	中国东北样带
278	NEE	net ecosystem CO_2 exchange	生态系统净二氧化碳交换量
279	NEEM	North Greenland Eemian Ice Core Drilling Project	格陵兰北部埃姆冰芯计划
280	NEMS	Nimbus-E Microwave Spectrometer	云雨5号微波频谱仪
281	NEP	net ecosystem production	净生态系统生产力
282	NETD	Noise Equivalent Temperature Difference	噪声等效温差
283	NIR	National Inventory Report	国家清单报告
284	NMHCs	non-methane hydrocarbons	非甲烷烃
285	NMO	normal move-out	动校正
286	NNSS	Navy Navigation Satellite System	海军导航卫星系统
287	NOSS	National Oceanic Satellite System	国家海洋卫星系统
288	NOSS	Navy Ocean Surveillance Satellite	海军海洋监视卫星
289	NPOESS	National Polar-orbiting Operational Environmental Satellite System	美国环境监测业务极轨卫星系统
290	NPP	net primary productivity	净初级生产力
291	NSCAT	NASA Scatterometer	美国国家航空航天局散射计
292	NSTEC	North South Transect of East China	中国东部南北样带
293	Oceansat	Ocean Satellite	印度海洋卫星

续表

序号	缩略语	英文	中文
294	ODP	Ocean Drilling Program	海洋钻探计划
295	ODP	ozone depleting potential	臭氧耗减势
296	OGO	Orbiting Geophysical Observatory	轨道地球物理观测
297	OPEC	open-path eddy covariance	开路涡度相关
298	OSL dating	optically stimulated luminescence dating	光释光测年
299	OTF	optical transfer function	光学传递函数
300	PAGES	Past Global Changes	过去全球变化研究计划（IGBP的核心计划之一）
301	PAH	polycyclic aromatic hydrocarbon	多环芳烃
302	paleo-ELA	paleo-equilibrium line altitude	古平衡线高度
303	PALSAR	Phased Array type L-band Synthetic Aperture Radar	相控阵型L波段合成孔径雷达
304	PAN	peroxyacetyl nitrate	硝酸过氧化乙酰
305	PAR	photosynthetically active radiation	光合有效辐射
306	PBL	planetary boundary layer	行星边界层
307	PCA	principle component analysis	主成分分析
308	PCBs	polychlorinated biphenyls	多氯联苯
309	PCMDI	Program for Climate Model Diagnosis and Intercomparison	气候模式诊断和比较计划
310	PDB	Pee Dee Belemnite	美国南卡罗来纳州白垩系皮狄组的美洲拟箭石PDB国际碳同位素标准
311	PDF	probability density function	概率密度函数

续表

序号	缩略语	英文	中文
312	PDO	Pacific Decadal Oscillation	太平洋年代际涛动
313	PEOLE	Preliminaire EOLE	"佩奥利"气象卫星
314	PE	potential energy	势能
315	pH	hydrogen ion concentration	pH 值
316	PLMR	Polarimetric L-band Multibeam Radiometer	全极化多波束 L 波段辐射计
317	PMC	polar mesospheric clouds	极地中间层云
318	PMP	probable maximum precipitation	可能最大降水量
319	POPs	persistent organic pollutants	持久性有机污染物
320	ppb	parts per billion	十亿分率
321	PPFD	photosynthetically active photon flux density	光量子通量密度
322	ppm	parts per million	百万分率
323	PP	precautionary principle	预防原则
324	ppt	parts per trillion	兆分率
325	PRA	participatory rural appraisal	参与式农村评估
326	PRC	power reflection coefficient	功率反射系数
327	PRISM	Panchromatic Remote-sensing Instrument for Stereo Mapping	立体制图全色遥感器
328	PSM	policy selection matrix	政策选择矩阵
329	PYRN	Permafrost Young Researchers Network	多年冻土学青年研究者网络计划
330	QNNP	Qomolangma National Nature Preserve	珠穆朗玛国家级自然保护区
331	QuikScat	Quick Scatterometer	美国国家航空航天局快速散射计

续表

序号	缩略语	英文	中文
332	RADARSAT	Radar Satellite	加拿大雷达卫星
333	RA	risk assessment	风险评估
334	REE	rare earth element	稀土元素
335	RES	radio echo-sounding	无线电回波探测
336	RM	risk management	风险管理
337	SAGE	Stratospheric Aerosol and Gas Experiment	平流层气溶胶和气体实验
338	SAR	Synthetic Aperture Radar	合成孔径雷达
339	SASS	Seasat-A Satellite Scatterometer	海洋卫星散射计
340	SAUS	shipboard ice alert and monitoring system	船载冰情报警与监控系统
341	SCAMS	Scanning Microwave Spectrometer	扫描微波频谱仪
342	SCF	snow cover fraction	积雪覆盖率
343	SCGCT	static chamber-gas chromatographic techniques	静态箱—气相色谱观测系统
344	SDAS	Satellite Data Acquisition System	卫星数据获取系统
345	SEARCH	US Study of Environmental Arctic Change	美国北极环境变化研究计划
346	Seasat	Sea Satellite	（美国国家航空航天局）洋面风场探测卫星
347	SEO	Satellite for Earth Observation	地球观测卫星
348	SHOALS	Scanning Hydrographic Operational Airborne Lidar Survey	扫描式水道测量机载激光雷达系统
349	SIFE	spectrally intergrated flux extinction	谱综合通量衰减

续表

序号	缩略语	英文	中文
350	SIR	Shuttle Imaging Radar	航天飞机成像雷达
351	SISA	spectrally intergrated snow albedo	雪面谱综合反照率
352	SLAR	Side-Looking Airborne Radar	机载侧视雷达
353	SLC	single look complex	单视复数据
354	SLR	Satellite Laser Ranging	卫星激光测距
355	SME	Solar Mesosphere Explorer	太阳－地球中间层大气卫星
356	SMMR	Scanning Multichannel Microwave Radiometer	多通道微波扫描辐射计
357	SMOS	Soil Moisture and Ocean Salinity	土壤水分与海洋盐分卫星
358	SMOW	Standard Mean Ocean Water	标准平均大洋水（国际氢氧同位素标准）
359	SMS	Synchronous Meteorotogical Satellite	同步气象卫星
360	SNODAS	Snow Data Assimilation System	积雪数据同化系统
361	SnowMIP	Snow Model Intercomparison Project	雪模式对比计划
362	SNR	Signal to Noise Ration	信噪比
363	SO	Southern Oscillation	南方涛动
364	SPAC	soil-plant-atmosphere continuum	土壤－植物－大气连续体
365	SPOT	Satellite Probatoire d'Observation de la Terrestre	斯波特卫星
366	SRES	special report on emissions scenarios	排放情景特别报告
367	SRTM	Shuttle Radar Topography Mission	航天雷达地形测绘计划
368	SSA	specific surface area	比表面积
369	SSCCC	spin-scan cloud cover camera	自旋扫描云覆盖相机

续表

序号	缩略语	英文	中文
370	SSM/I	Special Sensor Microwave/Imager	专用微波成像辐射计
371	SSRQTP	Scientific Survey and Research of the Qinghai-Tibet Plateau	青藏高原科学考察计划
372	SST	Satellite-to-Satellite Tracking	星间定位技术
373	SST	sea surface temperature	海面温度
374	START	Global Change System for Analysis Research and Training	全球变化分析研究和培训计划
375	SVAT	soil-vegetation-atmosphere transfer	土壤—植被—大气传输
376	SWE	snow water equivalent	雪水当量
377	SWS	snow water sensor	雪水当量仪
378	TCL	turbulence condensation level	湍流凝结高度
379	TCN	Terrestrial in situ Cosmogenic Nuclide	陆地原位宇宙成因核素
380	TCN	Track-Cross Track-Normal reference system	雷达坐标系统(航向-交叉向-法向)
381	TDRSS	Track and Data Relay Satellite System	跟踪和数据中继卫星系统
382	TDR	Time domain reflectometry	时域反射仪
383	Terra	Terra satellite	泰勒卫星(美国)
384	TIN	triangular irregular network	不规则三角形网络
385	TIROS	Television and Infra-Red Observation Satellite	"泰罗斯"卫星,电视和红外观测卫星
386	TL dating	thermoluminescence dating	热释光测年
387	TM	Thematic Mapper	专题制图仪(美国)
388	TOA	top of atmosphere	大气层顶
389	TOMS	Total Ozone Mapping Spectrometer	臭氧全量图分光计

续表

序号	缩略语	英文	中文
390	TOS	TIROS Operational Satellite	"泰罗斯"业务卫星
391	TPE	Third Pole Environment	"第三极环境"科学计划
392	TPO	Tree Preservation Order	护林法案
393	TSP	total suspended particles	总悬浮颗粒物
394	UARS	Upper Atmosphere Research Satellite	高层大气研究卫星
395	UCC	upper continental crust	上地壳
396	UHF	ultra high frequency	特高频
397	UNCED	United Nations Conference on Environment and Development	联合国环境与发展大会
398	UNCHE	United Nations Conference on the Human Environment	联合国人类环境会议
399	UNDRO	United Nations Disaster Relief Office	联合国救灾处
400	UNFCCC	United Nations Framework Convention on Climate Change	联合国气候变化框架公约
401	UPS	Universal Polar Stereographic projection	通用极球面投影
402	UTC	coordinated universal time	世界标准时
403	UTM	Universal Transverse Mercator projection	通用横轴墨卡托投影
404	UV	ultraviolet	紫外线
405	VHRR	Very High Resolution Radiometer	甚高分辨率辐射仪
406	VIRR	Visible and Infra Red Radiometer	(风云三号卫星的)可见光红外扫描辐射计
407	VIR	Visible and Infrared	可见光和红外
408	VISSR	Visible/Infrared Spin Scan Radiometer	可见光/红外自旋扫描辐射仪

续表

序号	缩略语	英文	中文
409	VLBI	very long baseline interferometry	甚长基线干涉测量
410	VOCs	volatile organic compounds	挥发性有机化合物
411	VOIR	Venus orbiting imaging radar	金星轨道成像雷达
412	WAIS	West Antarctic Ice Sheet	西南极冰盖
413	WCRP	World Climate Research Programme	世界气候研究计划
414	WGS	World Geodetic System	世界大地测量坐标系统
415	WHYCOS	World Hydrological Cycle Observing System	世界水文循环观测系统
416	WiDAS	Wide-angel infrared Dual-mode Line/area Array Scanner	机载红外广角双模式成像仪
417	WIGOS	WMO Integrated Global Observing System	世界气象组织集成全球观测系统
418	WMO	World Meteorological Organization	世界气象组织
419	WSN	wireless sensor network	无线传感器网络
420	WSS	Wide Swath SLC	宽幅单视复数图像
421	WUE	water use efficiency	水分利用效率
422	YD	Younger Dryas	新仙女木